Where The Crow Flies Backwards

Where The Crow Flies Backwards

Life on the Land in Aus

Douglas Cornwall

Title: Where the Crow Flies Backwards
First published in 2025 by Kani Consultants, Newcastle, Australia.

© Copyright on the text belongs to the author.
© Copyright on the compilation belongs to Kani Consultants.
All rights reserved.

No part of this publication may be reproduced, stored in a retrieval system, or transmitted in any form by any means electronic, mechanical, photocopying, recording or otherwise without the prior consent of the publisher.

A catalogue record for this work is available from the National Library of Australia

ISBN: 978-0-6450773-6-0
Where the Crow Flies Backwards / Life on the Land in Aus (through the eyes of an Irishman), Douglas Cornwall

Cover and illustrations by Leanen Deering

Keywords: Wit and humour, Agriculture, Adult humour, Young Adult Fiction, Australia, Animals & Nature, Rural Australia 20th century, farming families, anthology, short stories and jokes, Australian humour

Watch out for more books by the Irish immigrant known as Douglas Cornwall.

An insight into the author's mind

*sometimes called a preview,
foreword or introduction*

Young Douglas, as he is called, transitioned into an Ocker Aussie better than most. He maintained the Irish humour while collecting Aussie yarns. Then he decided he was going to share 100 of them.

Here are a few of the many experiences and observations of Aussie farming, including quite a few from his many acquaintences.

The one he enjoyed retelling the most is *The Two-Stroke Toilet*. Doug really felt sorry for that bloke.

Where the Crow Flies Backwards is a reflection of the way Douglas Cornwall sees life: spotting the unusual in what others see as normal.

At times you may need an Aussie dictionary. At the back you'll find 'definitions' for some of the weird and wonderful words we use in this great land.

But is it true?

All of these stories are true – in some degree, in one way or another. At least, this is the way Doug remembers them being told to him by … someone.

Granted, most of them were not created by him, many have been told to him or he overheard them. However, he may have been the first to record some of them, and he's definitely put his own spin on others.

If you find a story in here and believe you were the first person to record it and it's yours,

I'm sure he'll buy you a coffee and discuss the issue with you.

But, are these stories true? In today's age of questioning reality, history and memory, who decides what's true? And does it matter? You have to put the information into perspective, and read them in their context, and check their relevance, and then consider: whose point of view is it anyway?

Will it offend?

Some of these tales may not appeal to everyone. Some have the occasional swearword. And there is the occasional harmed animal (or human) – but never was anything damaged, harmed or injured for the purpose of writing these stories or publishing this book. The storyteller merely collated the facts afterwards – no different to a police officer or a paramedic really.

Australia is a multicultural country. So, when the author refers to characters by their culture or nationality, it is no different to mentioning their gender, age or hair colour. It is merely a way to help the reader imagine them. No malice or prejudice is intended.

Please enjoy exploring life on the land in Aus through the eyes of this ten-pound pom.

Or don't.

Cheers
Sandi,
An English Aussie with an Irish nanna.

Where the Crow Flies Backwards

Douglas Cornwall

Part 1: Nature
  Where the crow flies backward
  You know the wind is strong,

Part 2: People
  Where the people hurt and cry
  You know, well, that's just wrong.

Part 3: Truth
  When our words get all mixed up
  We can't tell dark from light,

Part 4: Morals
  When morals are in disbelief,
  It's time to put things right.

x

Contents

Part 1: Nature

Shut Up; He Pays Well	3
Poor Dog!	6
Ned's Weed Control	9
Well-trained Dog	11
The Two-Stroke Toilet	12
A Dog Named Blacksmith	19
Professional Hunters	20
I Think I'll Grow Chooks	23
My Aurukun Days	26
Next Door to Mooki	29
Grandad's Milk Cart	34
Neil was a Very Successful Pig Farmer	36
Strange Things Happen in Aus #1	43
The Dry Bauxite Mine	48

Part 2: People

Spinster from Paradise Station	59
Christmas at Cheesering	61
The Butcher's Name was Mr Push	64
Long Distance Vision	66
Dad's Old Station Wagon	68
Fokker Friendship	72
He Saved Lives by Doing Nothing	74
Just wanting to help	81
Charlie's Toilet	82
Freight On!	85
A Beekeeper in Trouble	88
The Worm Song	91
Can we help you?	92
Japanese Tourists	97

Contents continued

Part 3: Truth
The Bendemeer Bridge	105
Dad's New Tractor	107
I'm Nott, the Rabbit Inspector	109
Stuck – Out Here	111
I Don't Understand	113
White Fella Fishin'	117
Nan's Taxi Service	119
Strange Things Happen in Aus #2	122
Strange Things Happen in Aus #3	124

Part 4: Morals
Beekeepers can be an Ingenious Lot	129
The Town was in Chaos	134
We Were All Young Once, Okay?	138
Dad Wasn't So Sure	141
Camping Mishaps	144
Line Search Error	147
Signed, Fluor, Brown and Root	152

About the Author	159
About the Artist	160
Acknowledgements	161
Aussie Lingo	163

Part 1: Nature

**Where the crow flies backward
You know the wind is strong.**

Shut Up;
He Pays Well

Use your imagination and see eastern Australia, back before Captain Cook turned up.

The whole place was covered in trees. It would have been a beekeeper's paradise.

Now spare a moment for our brave pioneers who had to put a road through this formidable barrier. Armed with crosscut saw, axe, shovel, and crowbar, we owe them a lot. And then fancy feeding the men each day; how on earth did that happen?

So, it was in the early 20th century that an ambitious farmer bought a large block which straddled the Macintyre River in the state's north; covered in trees of course.

May I take it for granted that you are not volunteering to help me clear the trees off this two-thousand-acre lot?

You're wise; I'm dangerous with an axe!

Well, as it would happen, in those days a group of men were engaged to ringbark the trees in this chosen area near the river.

Once cleared, the ground would be good to run sheep on for wool, and they would have easy access to water.

They were shown the area he wanted cleared. "Don't go past that fence over there. Stay this side of this small hill, and clear up to that dirt road you can see from here."

They set to work with vigour and great intentions. The farmer boss would come by every couple of days to see the progress. He always came on horseback and followed the same track on the side of the river.

This was noticed by the group leader who instructed the 'boys' to change strategy, to which one of them complained.

"Shut up, let's just get this job done and we're outa here. He does pay well."

They agreed and work continued. They were good workers, and the boss was impressed by their progress.

I'm guessing, but let's say the whole job took three weeks. When the final tree was ringbarked, the leader turned up at the house, and the farmer boss made sure he and the men were paid the right amount.

They shook hands and the fellow disappeared.

Outa
A lazy way of saying "Out of", such as "I'm outa here"; not to be confused with outer which means the outside of something.

The boss farmer became quite busy with ploughing and repairing damaged fences. It took him a while to get back to the trees. He then had to organise another mob to push over the now-dead trees, and later to clear all the fallen timber to prepare the ground for grazing animals.

When he did get back – as usual, on horseback, following the same track on the side of the river – he couldn't figure why the trees weren't more dead than they should be.

A couple of weeks later, he rode over again: on horseback, following the same track. He discovered that, yes the trees were dying, sort of, but they should be completely dead by now.

He got off his horse and walked into the area of ringbarked trees.

Oh My Goodness. The men had only cut halfway around the girth of the trees! They had figured out just how much of each tree the boss could see from horseback and cut just that much. He was not happy, and he never trusted contractors again.

These early farmers worked hard and carried the nation's finance on their backs.

We were a wealthy country back then – praise to the men and women who made us what we are.

I feel a sermon coming on!!

Poor Dog!

In the days of the PMG (or Telecom) a phone techie took a call from a farmer out west.

"How come I don't receive calls, well very few at least? I've had people bail me up in town saying, 'We tried to call you, but you wouldn't answer', but I never hear nuffin. What's the point of having a phone and not hearing people trying to call you?"

PMG
The Postmaster-General's Department provided the postal and telegraphic services throughout Australia.

Telecom
Australia's earlier telecommunications systems

Nuffin'
A lazy way of saying "nothing".

The techie was asked to come and see what the problem was.

"Yes okay. Will next Thursday suit?"

Early on Thursday, our techie friend rang the farmer to make sure it was still okay to come out and see what could be done.

The farmer took the call and agreed that this afternoon would be fine.

"Before you go, you said you couldn't take calls; but you answered my call."

"Oh yes. You see, the dog barked!" The farmer answered confidently.

The techie thought to himself, *is the problem with the phone or the owner? What exactly am I getting into here?*

That afternoon, he arrived on the farm and was welcomed in.

"The phone's in the next room."

Our friend then approached the phone, checked the connection on the phone itself, and looked at the wall socket. Everything seemed fine.

He called the exchange who called him back – no problems so far.

"Where does the phone line come into your property?"

Techie
A lazy way of saying "Technician" which is someone with expert knowledge on the technical elements of a subject.

He was shown where, "I'll start looking from here."

"Yep, no problem. Come on dog, I need you in the sheep yards."

He disappeared, and our techie friend followed the line under the house.

Oh My Goodness.

The farmer was tying up the dog to the phone line so he could have the length of the house to run.

Over the years, the metal loop on the dog collar had worn through the insulation on the phone line which meant that, should the dog be actually tied up to the line when someone called, the poor animal would get an electrical shock. Hence the dog barking would tell the farmer he had a call!

Ned's Weed Control

Have you noticed that children have a habit of growing up and either leaving home or marrying? Or both. Eldest daughter finished school and married. Soon enough, Ned found himself to be a grandfather.

Daughter and son-in-law rented a small but modern house nearby and family life proceeded. At the rear of the house was a small well-kept lawn with some flower beds. Son-in-law maintained the small lawn and flower beds quite nicely and he was very proud of his gardening efforts.

To one side of the rear lawn stood the outside laundry and hot water system. The hot water system was only partially enclosed but was well shielded from the elements. It was a gas system but proved to be very effective, even in cold conditions.

Son-in-law was concerned at the ever-increasing weed growth, so our new grandfather volunteered his help. Being a successful farmer, he knew how to control weeds – the cheap way.

One day, son-in-law came home early to find his father-in-law weeding the footpath outside the laundry with his favourite weed killer: petrol.

"Dad!" exclaimed the proud gardener. "You can't use petrol that close to a gas hot water system!"

Dad was most indignant but when they were at dinner together, they all saw the funny side of things and the old farmer was quickly forgiven. The incident forgotten.

But …

A few years later, after father-in-law had learnt the necessity of using *normal* weed killer, the young family went on holidays.

By now it was expected that he keep the gardens looking nice and weed free. So, this time he got out the weed killer, one of the strongest types, and continued on his father-in-law duties.

Unfortunately for him, during his conscientious work with the weeds around the back lawn, he must have unknowingly formed a pool of the very strong weed killer; and stepped in it.

Days later, these strange footprints appeared on the green grass.

Patches were dying in the shape of large feet.

Several weeks later, son in-law was still laughing at the situation when he told my wife and I about it.

Father-in-law was a very successful pig farmer – but not so successful at weed control.

Well-trained Dog

A professional duck hunter trained his dog to retrieve a downed duck without biting it.

What's more, he could retrieve it over any terrain, especially water. The dog could even bring in a downed bird by walking on water.

One time, the fellow took along a close friend and they downed quite a few birds. Each time, the dog got the bird and walked on the water to bring it in.

On their way home, the proud owner of the retriever asked his friend what he thought of his dog. "Did you notice anything clever about him?"

"Yes – he can't swim!"

The Two-Stroke Toilet

Once upon a time long long ago (yeah yeah, we've heard it all before)...

I was travelling westward from Croppa Creek to a farm off the Newell Highway, because I'd been asked to work on a tractor's air-conditioning system.

I arrived early, just in time to watch the farmer manoeuvre his many machines in and out of his machinery shed so as to have this tractor close to a power point.

We were standing next to his grain truck when a yellow harvester came through his front gate and found its way to us.

We had no inkling of the emergency in the driver's cabin.

The machine stopped, the cabin door flew opened, and a figure leapt to the ground. The driver, not knowing which of us was the owner of the property, ran up to me and demanded directions.

> **Yeah**
> A lazy way of saying "yes".

"Quick – where's the toilet???"

We now knew what the emergency was.

Obviously, the emergency first occurred somewhere on the Newell Highway. There are very few toilets on the Newell and the one that did exist got burnt down – there is no justice in life, is there?

The driver, having spotted the building near the orchard, was already on his way.

"No no!" The farmer sprang to life. "Over here; to your left!"

Now, you and I know what a toilet looks like – am I right? But this building had been built by an industrious farmer on the western plains.

The building was twice as wide as a normal Aussie dunny and was known as a long-drop. I'm guessing you don't know what a long-drop is.

You see, out here there is an abundance of artesian (underground) water. It can be deep down in the earth's crust.

Dunny
an outside toilet, usually away from the house

Long-drop
a pit toilet

Artesian
underground water supply

When you've had a driller find water for you, it is normal to ask the operator to drill a hole (somewhere near the house or shearer's quarters) which can be used as a toilet. Naturally, they just drill down about ten or twenty metres, not going anywhere near the artesian water table, and there you have it.

So, with a small building, a door and a newly constructed seat, you have a very effective waterless toilet that will last, not years, but two or three generations.

BUT ...

Houston we have a problem here!

You see our 'farmer of the year' had built a two-seater bench (with a dividing wall for privacy of course) should two people wish to go at the same time. What with seasonal fruit pickers, shearers, contractors and such like, this is perfectly reasonable.

But, on this occasion, the two-seater with a short partition came close to causing an anatomical disaster!

Our short tempered contractor, having spotted the unusually wide toilet, made a beeline for the toilet door.

He had already started undressing. He entered, disrobed, pirouetted, and sank down onto the seat.

Stunted anatomical processes had begun – at last. Life had meaning again!!

BUT WAIT.

"What the hell is that noise?" He asked himself as he realised there was something hostile just below his bare ... how shall I put it ... backside.

Anatomical processes screeched to a halt – now *that* takes so much mental effort.

He raised his ... what shall I call it ... from the seat and listened to this totally unexpected hellish noise that he instinctively knew threatened his ... what shall we call it??

His mind froze at the awareness that the noise came from a thousand or more flesh-eating, germ-infested, disgustingly HUNGRY mosquitoes.

A poorly dressed figure appeared at the toilet door. The farmer and I froze at the sight. He was obviously in trouble!!

"The toilet is full of mozzies!" He screeched.

Our farmer, seized by the importance of the moment, shouted in reply, "WAIT!"

He didn't just run, he supercharged into the machinery shed and quickly returned and headed for the distressed contractor.

With the smallish petrol can in hand, he yelled apologetically as he raced to Ground Zero, "I've been meaning to do this for weeks now".

Mozzies
mosquitoes, a much-hated insect.

The victim of the moment, still hanging onto his half-mast trousers, stepped away from the door – not too sure who was being attacked here, himself or the mozzies.

Into the toilet our farmer friend went, ripped off the petrol can lid, and poured about half a litre of petrol into the toilet internals.

He threw in a lit match, and waited for a reasonable 'whompa' from the long-drop before he exited the building and announced, "They're gone now."

During this short drama, the contractor with the cruelly halted 'anatomical process' was near panic.

Look at it from his point of view.

The nearby orchard presented a place of relief and solitude – something sadly lacking since this crisis began. But this wasn't his farm, and he knew the farmer was genuinely trying to help.

On hearing the words, 'they're gone now', he levitated back to the seat and downing his trousers, he sank back onto the toilet seat.

'Full steam ahead' his brain signalled to his digestive tract. Pent up anatomical processes resumed.

Whompa
a dense sound made by machinery such as helicopter blades, or a fire in a small area.

But wait!

"What the hell is that new noise???" Gone was the near-lethal buzz of a thousand flesh hungry ... you know what I mean.

Now we had a slowly pulsing whompa whompa noise. It wasn't a continuous, smothered explosion, just a spaced whompa (two seconds) whompa (two seconds) whompa ...

The possessor of the newly halted anatomical processes was in near revolt mode. Remember, his poor digestive tract had been given the go ahead. His internals had begun the 'download' shall we say – but his tortured mind had commanded STOP. Again!

Back to the whompa whompa whompa. His frenzied thinking had not yet grasped what the problem was but demanded action.

What had happened was this: Our friendly 'farmer of the year' had set fire to the contents of the long-drop. The mozzies were certainly history, but we now had an oxygen-hungry fire looking for an exit from this claustrophobic cavern in the earth's crust. Having found oxygen, the flames would reach for the second hole in the seat above.

The first hole had mysteriously been blocked by a 'you know what' – a very bare and trembling human rear end.

The whompa noise came from the flame reaching up and out of the second toilet hole. Having consumed the available oxygen, it would die back. Finding more oxygen, another whompa and another flame protruding from the second seat hole, only to die back again.

Remember, two seconds between every whompa whompa.

It is here that our innocent contractor friend (whose bodily functions will probably never be the same again) came to realise what was happening was on the other side of the partition.

Now look, we all make mistakes.

His frozen mind was still inquisitive enough to wonder what this unearthly whompa whompa was just the other side of the short partition.

So, he gingerly lifted off his toilet seat and peered around the partition at the second hole next door.

Mistake. BIG mistake.

The oxygen-hungry flames, on finding a new exit, shot up and around the human form. Every hair on his 'nether regions' was instantly and wickedly singed.

Having a scorching flame ascend from down there up between your legs and towards your chest is the stuff of nightmares for us blokes.

The toilet door burst open! An indecently clad figure threw himself out onto the ground. He jumped up, clutched his trousers and headed for the orchard. I assume to resume normal bodily functions.

Who needs toilet paper in such an emergency?

A Dog Named Blacksmith

A city chap wanted to visit his country cousin. He arranged to stay for a week and on this day, he found himself helping with working the sheep.

During the course of the morning, he learned that the name of the sheep dog was Blacksmith.

Over lunch he asked, "Why do you call your sheep dog Blacksmith?"

"Oh that's easy. When he hears a gunshot, he makes a bolt for the door."

Professional Hunters

Here's a true story my aboriginal helpers told to me with quite some delight.

Whenever the Aurukun menfolk felt the need for eating meat, they would go hunting pigs.

These men had a favourite strategy using the local topography to their advantage. They would deliberately drive a mob of pigs into a peninsular from which they could not escape easily.

This peninsular had a ridge roughly up the middle of it which meant they could divide the gathering hunters into two groups. One would drive some of the mob up the right-hand side, and the rest could drive the remaining ones up the left-hand side.

> **Menfolk**
> the male members of a
> community or family.

This always worked out so the sheer retelling of this tale brought smiles to their faces.

On one occasion, things didn't go as planned like it always had in the past. The whole point of using the peninsular was because peninsulas always come to a point – in this case, the Archer River.

Here, the mob would be trapped between the river and the men with spears, rifles and whatnot. They would gather the mob into a small, cleared area and pick off whatever would keep themselves and their families in meat for a few days.

I ask you: If these men didn't deserve the title of Professional Hunters, just who did? The Wik men and women knew all about catching and preparing meat, which was often cooked in a ground oven. I was lucky enough to have see this done.

Sorry, back to the task of picking off an animal they want.

It just so happened, on this one occasion, that one group of men arrived at the clearing first with the other group way behind.

Whatnot
an insignificant item

Pick off
select one by one.

21

The tellers of this story were trying not to enjoy the recalling of this.

When the second group approached the clearing, they expected to see their friend's corralling and trying to get a good shot at a pig. There wasn't a pig in sight!!

What the ... ???

All they could see was the weaponry the first group had, but it was all scattered on the ground. Not a pig or person in sight!!!

Their astonishment was complete.

What the ... ???

Then they heard from up in the trees above: "Look out behind you!"

They didn't look. They joined them. Up in the trees. Far off the ground.

You see, what the second group didn't know was that they were being followed by a large wild pig. It was nearly the size of a cow, with large tusks, ready to do some damage to anybody who got in its way!

That's all I can recall of the story, but I can still hear them laughing as they retold it to me.

I Think I'll Grow Chooks

Luigi was a coalminer. He'd worked in the coalmines since his teenage years.

But one day, when he received his pay, there was an important document with it. A redundancy notice.

He went home after work and broke the news to his wife.

"Well love," said he in that high pitched mediterranean voice. "What are we gonna do now?"

They had three months to think about it so they made, and enjoyed, a cuppa.

The day came when Luigi walked out of the mine for the last time. When he got home, he announced: "Well darling, I've made a decision. After all, I came from a farming family so ... I think we'll grow chooks."

Cuppa
a cup of tea or coffee

They sold up in the Hunter Valley and bought a place somewhere near Tamworth.

Luigi set about creating a chook farm and, when he had things ready, he headed into town and showed up at the local Poultry Coooperative or 'The Co-op' as the locals called it.

"Yes sir, what would you like?" the friendly young salesman asked.

"I'll take a thousand day-old chicks please."

"Yep, okay, but I couldn't have them ready for you before next Wednesday."

"That's okay."

The next week, Luigi drove back in and picked up his a thousand day-old chicks and headed for his farm.

About a week and a half later, Luigi turned up at the co-op again. By now, they knew who he was.

"Yes Luigi, what would you like?"

"I'll take a thousand day-old chicks please."

"Again? Yes righto. Is next Wednesday okay?"

"Yep, that's okay."

Again, he waited a week, then picked up his chicks and headed home.

Nothing's seen of Luigi for nearly a fortnight and the salesman was just thinking about how the newcomer is going with so many chicks, when, once again, Luigi fronted up.

"Hi Luigi, how have you been, mate? What would you like this time?"

"I'll take another thousand day-old chicks please."

"Now hang on Luigi, these are big orders. What's going on?"

"I dunno," says he, scratching under his hat. "Maybe I plant them too deep!"

Just in case you were worried, this is not a true story.

Dunno
a lazy way of saying "don't know"

My Aurukun Days

During my stay at the North Queensland town of Aurukun, I saw things that were very unusual.

The founders of the village had provided a primary school and a domestic science classroom which the local youngsters attended and learned skills they could use when looking for work away from the village environment.

This classroom was well attended but, occasionally, there developed a problem.

You see, Aurukun is on the west coast of the Cape York Peninsular right in the path of the yearly cyclones.

The domestic science classroom was self-standing, and the upper half of the tall walls were all louvres.

Over the years, louvres were either lost or broken which, in non-cyclone country, isn't a problem. But this is North Queensland; cyclone country, and it is a problem, believe you me.

One of the rules for surviving a direct hit by a cyclone is to ensure that the huge winds don't get into your building and under your roof. Believe me, it does happen.

So, how do you prevent the damaging winds from getting in and wrecking the building when a fair percentage of the louvres are missing?

It was quite amusing to watch this.

Let's say the cyclone is coming from the east. The Bureau of Meteorology (The BOM) have been warning you for days now, and it's definitely coming, and getting closer by the hour.

You muster up a few volunteers and start removing louvres from the west side wall and fill in the eastern wall. You keep at it until the whole wall is complete with no gaps.

The next step of survival is to watch for any problems while you stay in the room; keeping your head down of course.

The cyclone is overhead now and there is an uneasy calm.

"You beauty! Let's rattle up some tea and scones. What on earth are domestic science rooms for anyway??"

Not blooming likely.

As the cyclone continues on its western course, the winds switch around and come from the west. You have just removed a hundred louvres from that wall to fill the eastern wall!

This is the funny bit for onlookers!

You now have to remove the louvres from the eastern wall and fill up the western wall.

And you have about fifteen minutes to do it!

PANIC STATIONS.

You watch incredulously as the small army of people hurriedly remove louvres, run over to the other wall and fill it up.

It's a routine they are quite used to and do not take kindly to any mirth on your part!

Trust me, I know.

Next Door to Mooki

Way back in the mists of time; let's say in the late 1950's, successful farmer Charlie decided to sell up, somewhere near Wang I think, and leave Victoria.

He and his family decided that northern New South Wales would suit them both, weather-wise and economy-wise.

With a sizable bank balance, and after some inspections, he settled on a farm in the Inverell area.

To the naïve non-local, it seemed surprisingly cheap but he and his wife soon began the process of establishing a new home and getting the farm productive again.

Eventually, he discovered that he had paid well above the average price per acre for land in the Inverell area at that time. Nevertheless, under Charlie's industrious hands, life became very productive, the farm grew and the family prospered.

Charlie had a wide sweep of knowledge and skills. He was competent at building sheds, repairing tractors, working with livestock and tending to them when they were injured or sick.

I watched him one day when a cow was having a breach birth. He was able to push the partially showing calf back into the uterus, turn it around, and assist the mother to birth the calf normally.

To a farmer, losing an animal is serious business. I remember wondering: who is the most relieved, mumma cow or Charlie?

The years rolled by, and the children were in high school. Charlie wished to expand his enterprise and was offered a large farm out in the Tingha bush. And I mean BUSH.

This farm was originally created and settled by two brothers, before the first world war. They built their own home out of timber they felled themselves. The roof was a shingle roof, the floor was just the dirt the house was built on, and the small kitchen featured the wood-fired stove where the large teapot was always on the fire.

If you've never tasted tea from one of these early day always-on-the-fire teapots, you may consider the gods have been kind to you. "Why?" I hear you ask. After a few days of being on or near the fire, the liquid is more like bitumen than pleasant tea.

The two brothers were skilful farmers who had the main building complete. They had the shingle roof as rainproof as possible, the vertical poles forming the walls were sturdy, but the gaps between them allowed a considerable amount of draught through.

How were they going to seal the gaps to make the place truly livable?

I'll tell you. With newspaper. Back in the days of print, newspaper was plentiful. And the glue was made from kitchen products. Simply apply a liberal coating to each log and apply one sheet at a time until the whole wall is covered and draughtproof. That's how.

I well remember standing still for many minutes and reading 1930's and 1950's newspapers.

Let's back up a little here, remember he found out too late that he had paid well over the current price per acre for his other farm in Inverell? Now that cattle prices were on the up and up, he was contemplating a second farm to extend his ability to have more livestock.

So he bought the place near Tingha and began the process again, re-vitalising an old farm that hadn't been used for some time.

One of the first important jobs is to get a clear mental picture of where your boundary fences are. In this case the neighbouring property was owned by a Dennis Cleary.

So, by telephone they arranged to meet. I was included and, on horseback, the three of us met at an old partially-burnt building known as Charcoal Inn. We crossed the Moredun Creek and headed for the nearest boundary fence. Dennis led the way.

Pointing to a fence he declared, "This side is yours, and the other side is mine, called Mooki". We did this for a few hours, and you must remember the area was only sparsely cleared and we needed to know where the boundary fence and where the corners were.

The whole time I rode behind the two, listening and watching. This is where I get near the climax of this story. I could tell that Dennis was asking questions that were leading up to something, but I couldn't figure out what he was getting at.

Finally, he wanted to show us where his house was, so we went through a boundary fence gate and entered Mooki.

We were nearing his cattle yards near his house when he asked, "Charlie, guess how far down into the ground I put the yard posts?"

"I have no idea, Dennis."

"I dug them down to six feet." But that wasn't what he really wanted to discuss.

It was here that he finally got the courage to ask what he had been wanting to know all along.

"Say, Charlie – how much did you pay for the Caley brothers' old farm?"

I knew instantly that this was the real question he'd been burning to ask, but he wasn't prepared for the answer.

"Three pounds an acre!"

Dennis didn't hide his shock. He spun around in his saddle and almost fell off! *You try spinning round in a saddle next time you're in one – you'll see why I gasped and smothered a laugh.*

Dennis' face looked as it was draining away from his head. He might have even stopped breathing for a few seconds.

You see, this was *below half* the price agents were asking for this type of land.

Secretly, I know that Charlie had a meeting with his Creator during a Billy Graham crusade and I'm guessing his creator had 'balanced the books' for him.

Grandad's Milk Cart

A young son, home from college in Dublin, was riding on the back of his father's milk cart. They were on their way to deliver milk to his father's customers. Just as he did when he was young, the son helped his father whenever he returned home.

In those days, the milk farmer would milk his cows and pour the milk into a milk urn, sometimes called a churn.

It was the son's job to dip the specially designed ladle (holding about one pint of milk) into the churn, filling it with milk. He would then walk up to the front door of a house and pour the contents into the milk container left out by the customer. He would pick up the small amount of change and return to the cart.

Dad would put the money in his pocket, and off they would go to the next customer's house.

But the son had grown up.

The son, having learnt economics at college, got to thinking that his dad spent a lot of time on the cart being towed by a goat.

"Dad."

"Yes son?"

"Wouldn't you be able to spend more time attending to crops and gardening your vegetables for market if, instead of using a goat to pull the cart, you got a pony for the job?"

Father was silent.

"But don't you see, Dad, the pony would be much quicker, and you'd have more time with all the other things you do to produce an income?"

"Um," said father.

The son persisted, but his father kept quiet for some time.

Father then spoke. "Look son, I don't doubt what you're saying, but I'm pretty sure the pony wouldn't eat the thistles."

The son was speechless at such logic.

Neil was a Very Successful Pig Farmer

True story. Let me tell you about a pig farmer of repute.

Somewhere near Tamworth, Neil ran a very successful piggery, but he also started a fruit orchard. Two incomes are better than one, he thought, especially when you're farming.

As the years rolled by, Neil extended the orchard a little at a time – when the pigs allowed. Bit by bit, the orchard increased in size each year and proved its worth when market prices went against pig farming.

The time came when the fruit harvest was too much for Neil and his wife, so fruit pickers were hired annually.

No problems yet.

One day, when his children were home on school holidays, they got busy exploring the garage. Neil was busy with the pigs when they showed up, eldest in the lead, with a very old biscuit tin.

"Dad, what's this?"

Dad, being annoyed at the disturbance, only planned to take a quick look inside the open biscuit tin his daughter was holding.

He momentarily froze in horror. Then, quickly recovering his composure, he seized the tin from the child's grasp and said, "Go to your mother now!"

They did, not sure why, but when dad spoke like that, there's no arguing.

So, there he stood, frozen to the spot. The pigs were going to have to wait.

Maybe I'm having myself on here, he thought to himself.

Gingerly he opened the lid. Yep, sure enough, in the tin were three, maybe four, sticks of gelignite.

Holy cow, what am I gonna do with this stuff?

Let's call it 'real estate enhancement accelerant.' There's enough in here to put a fair size hole in anybody's back yard. Not to mention the neighbour's back yard, and the back of their house too.

His brain scrambled to think.

What in hell's name could I do with this stuff? I know, I'll put it down the well!

Gonna or gunna
A lazy way to say "going to". A person who intends to do something but never gets around to it.

This idea didn't last long as he realised that the drop would be enough to blow himself sky high.

You see, the trouble with this type of explosive is when it ages it sweats. Little beads of yellow stuff form on the outside of the stick. It is said that the beads themselves are a form of nitro-glycerine. This stuff only needs the slightest jar and up she goes.

Now you know why dad had the horrors at the thought of the eldest child tripping and dropping the tin.

Walking away from the well, he thought, *Ahha! I'll put in the loft of the machinery shed.* Nah.

Ah right, I'll carry it down to the far end of the orchard and bury it.

He did just that. You see, to carry it on a swaying tractor would be too risky, so carry it he did.

On returning to the machinery shed, he loaded shovels and crowbar onto the tractor and proceeded to the chosen spot on the far side of the last row of trees.

"By the time I'm finished here," said he aloud, "the kids will never find it."

He had picked a spot where there used to be an old well, gently lowered the 'real estate enhancement accelerant' by rope, he let it settle on the bottom.

Next, spade by gentle spade full, he dribbled soil down onto the 'you know what'. Eventually, he could shovel in more dirt at a time. Having filled the hole, Neil used the tractor's front blade to level the ground.

"There. That should do it. The kids will never find it there."

They never did. But something else did!!

As I was saying: two incomes are better than one when you're farming. A lot of money was required to feed a family, pay the mortgages, and all the vet bills.

As the years went by, and the orchard was extended bit by bit each year, Neil came to rely on the fruit pickers to harvest the orchard fruit.

In this particular year, Neil was very pleased, with the approaching harvest looking like the best ever, so the same group of fruit pickers was contracted to do their work. They duly turned up and set to work. They had it down to a routine by now and their leader had a fair grasp of the English language.

This particular day started normally enough – as most catastrophic days do. The contractors were busy picking fruit, handling each piece with care, and Neil was busy working with the pigs.

Suddenly, a very frightened fruit picker burst through the end door of the pig shed and ran up to Neil.

Panic disrupted her normally beautiful face as she almost screamed the word "SNAKE".

Thinking Neil hadn't heard her, she repeated it, only this time she gestured with two fingers: 'Hole.'

By now Neil had grasped the urgency of the situation and asked, "Where?"

"You come. Come, I show you."

She sprinted for the door, Neil followed out into the pig yards, past the garage and into the orchard.

"By jingoes," said he to himself. "This lady can move."

About three-parts of the way down the line of fruit trees, she stopped. Looking around she was trying to find the 'hole' mentioned in the pig shed.

About here, Neil started to get a touch upset – there wasn't a fruit picker in sight.

"There. Hole. Snake go there! You get rid of snake please – very urgent." She disappeared as quickly as she appeared at the shed leaving Neil pondering the situation.

Sure enough, thought Neil, *I wonder what size animal made that hole.* It was certainly big enough for a snake to take up residence!

By jingoes
an exclamation expressing agreement

For a moment, he stood and wondered what to do.

Dear reader, please understand, our very successful pig farmer was one of those people who, once they have a plan in mind, will go through with it; sometimes without careful consideration of possible side effects.

What he hadn't realised is that the previous tenant of this hole – a rat or some other animal – had chosen this very spot because the dirt had been disturbed earlier making the job of creating a nice little underground hideaway for itself that much easier.

Are you with me here?

Retrieving a small tin of petrol, he proceeded to pour some of it into the snake's nice little hideaway through the hole that the fruit picker had observed the snake going into.

Next, he laid out a short length of rope he had brought with him and started to soak it in petrol to form a wick which he hoped would give him a short time to get away once it was lit.

After a moment's thought, he decided that it wasn't long enough, so he unplaited it, which made it three times longer.

Another soaking in petrol, and he fed it down into the snake's internals.

He took a quick look around to make sure no pesky fruitpicker was getting inquisitive. There was still none to be seen.

Little did he realise that this is the exact spot where, years earlier, he had buried the biscuit tin of very high 'real-estate enhancement accelerant' explosive stuff.

Glancing around again, he lit the end of the wick, turned and began to make great strides.

What he didn't take into account – but you and I can – is the flame didn't obediently burn the wick slowly. The little flame simply flashed along the top of the rope and straight down into the snake's internals.

Neil was planning a smallish explosion just enough to terminate the hiding inhabitants, but, what he got was a fantastic, near Richter scale happening.

Said he, when things had died down somewhat, "All I can remember is that I was running, and still in contact with the ground but I had gained about three feet in altitude."

Strange Things Happen in Aus. #1

True story.

A strong family in Western Australia were very successful at farming. Mum and dad were the farmers and in-laws helped as well as can be expected when you consider the size of some of these properties. They ran quite a large herd of cattle.

This part of Australia suits cattle farming. You need plenty of land and you need plenty of stock. You need plenty of water to sustain yourselves and plenty of luck and knowledge to be able to sell at the right time and price.

As can happen, mother nature will find a way for babies to be born, and so it was with this mum and dad. In fact, as the years wore on, this farming family had four boys.

Over time, each son was given their own job as they were big enough and strong enough to do them.

The years passed quickly and the eldest was due to finish high school the following week.

"Well son, what are you going to do with your life?"

I remember being asked this question. I just wanted to be a bee farmer. I never was, but that's a story for another day.

"Dad, I want to join you and mum on the land please." He was doing a fair bit of the tractor work anyway and wasn't expecting dad's reply.

"No son, I want you to go and learn a trade or some profession first!"

"What? Why?"

"Well, as you know, farming can be a risky business. I want you to have something to fall back on when farming goes against us!"

The eldest son was not pleased and began to argue. Mum usually tried to keep the peace, but knew she had to stay quiet or it would make things worse.

Dad won the day and the eldest son headed east to find what life had in store for him.

Well, you've guessed it. The second-born finished school and approached father, fearing the worst. He was right to do so!

"But Dad, all I know is farming!!"

"That's the whole idea son, you'll need something else to sustain you (and possibly a family) should our economy go belly up.

Yep, you're right again. Son number three finished school and had the same confrontation with dad. With the same result.

Well, not to labour the point, son number four finished his schooling. He didn't even bother fronting dad; he'd had his belongings packed for weeks.

Mum drove him to the travel agency, kissed him goodbye, and wished him luck.

There's something I haven't mentioned *(and I speak from personal experience).*

Young lads from off the land are in high demand with employers. They are quite used to working their socks off when there's lots of work coming in; they have spent years working as youngsters with animals and machinery, and repairing fences or broken cattle yard rails.

You name it, they've done it.

Unlike some of their city fellows, these guys are willing to work hard, and they have a vast range of knowledge. Farming dads make sure of this.

Belly-up
to go broke

Well, like I was saying, number four landed an apprenticeship as a plumber somewhere in Melbourne.

Son number four was popular with his employer right from the outset. He quickly picked up new skills and would tackle even the hardest of jobs with enthusiasm.

He was near the end of his time when a telegram arrived from WA. Naturally, it was delivered to the office not the workshop.

The office staff could see who it was meant for but could not understand the meaning of it.

It was handed to the big boss who looked at it – this way and that – but the words were totally meaningless.

"Okay. Give it to him, hopefully he can understand it."

One of the juniors brought the paper out to the work bench of son number four, and threw it down. "Telegram for you!"

Our hero from WA glanced at it, but was determined to finish off what he was doing.

Having finished, he picked up the telegram, read it, and stiffened with a frightened look on his face.

He torn off his work apron, ran out of the workshop and through the front office. As he passed the boss' office he half shouted, "My folks are in trouble boss! I'll be back!"

The boss wanted another look at the telegram because he'd never seen our hero panic before. He tried again to make sense of the message but couldn't.

The message read, 'No wind. No water. Send pumps'.

Son number four was the only one with the resources, the 'nous' (*pronounced as in house but without the h*) and the connections to help overcome the problem and get those thousands of head access to water.

In Australia, there was plenty of artesian (underground) water, and in the early days, windmills were commonly used to lift the water up to the surface.

It all works simply and beautifully until there's NO WIND.

The Dry Bauxite Mine

Water was never a problem at Aurukun except during the wet season. This produced several unusual happenings.

When I was still new there, every morning I would wonder about this strange routine. Several of the local men would drive our small Dodge truck up and down both airstrips, which formed a large T-intersection, and intently watch the ground.

Okay, it was my first wet season and there's a lot to learn, let me tell you. I refused to ask why – I didn't want to display my ignorance.

It just so happened that this is the time I began to have my quiet prayer time while walking along the airstrip.

Wet season
the period of heavy rainfall, usually
October to April in the
northern parts of Australia.

The solitude in the evening was wonderful with no one within coo'ee to cause distraction.

My mind was elsewhere when I heard this sucking sound – the same sound you hear in the laundry tub when you pull the plug. I won't even try to textualise it, just use your God-given imagination and see if you can work it out any quicker than I did.

I looked around for company, but no one was anywhere near me. I then looked at the ground around me, and there to my incredulous gaze was a hole in the ground off to my right. It wasn't there on my approach but it sure was now – about the size of a soccer ball.

Ahha. Now I know why the men patrolled both airstrips and would suddenly stop and start throwing largish rocks at the ground.

When I asked about this, they replied, "Because it's the wet season." One guy did explain a bit more: sometimes the ground will develop those holes as the saturated soil fails and the men have to fill them up

---

Coo'ee
a signal used in the bush to attract attention. The first syllable is elongated and the second is of a higher pitch.

Pothole
a hole in a road

before a plane comes in and puts a landing wheel in them. This will cause quite a sharp and unnecessary jolt which could damage the landing gear.

I'm used to potholes in the roads around Inverell, but this all seemed ridiculous.

"Crikey," said I.

Another strange effect of the wet season is the fact that nearly half of the surrounding bush country becomes inundated with water from the overflowing creeks and rivers.

This has the result of forcing a lot of land animals to congregate on the higher land. Can you imagine all those animals now trying to co-exist in something like half the thousands of hectares that they are used to.

If the animals could speak, they'd be saying, "Oh look, here's a lovely flat open area; let's live here till things dry out".

Let me tell you about the commercial airline pilots who were flying out of Cairns, doing regular stops on the Cape before flying into the NT. They would discover their power of speech when, on final approach, a mob of animals would run out onto the landing strip – and the language is not very civil either.

Crikey
an expression of surprise

To counter this frequent occurrence, they would approach the landing strip at above-normal landing speed. This would enable them to pull the plane back into the air and go around.

They were fantastic pilots.

But, Houston we now have a problem.

So, on this particular day, the pilot decided to stay and land as per normal but this is the wet season and things aren't normal, I can tell you.

We now have an aircraft on the ground but at a speed way above normal. Because the wheels are on the ground – the very wet ground – the brakes aren't working normally, and we have probably half the normal braking power.

The end of the strip is getting closer and we're still travelling too fast. I was lucky enough to see this – with his wheels still on the ground he or she would lift the aircraft's nose up and use the underbelly of the craft as a sail.

This works magnificently and the craft slows to a near halt before he/she turns it around and comes back to where the people are waiting.

Gotta
A lazy way of saying "got to" as in
"I've gotta go now".

You gotta see this to believe it.

The pilots of this airline that was based in Cairns, took their job quite seriously. Having landed the plane, they were used to waiting for the passengers to get off with their luggage.

They were also used to the European wives bailing them up and having a good old chin wag. This was the only way the wives could catch up with the outside world and would hold a conversation for half an hour at least.

These pilots knew how much the informal talking was appreciated and how much they knew that the folks 'down there' wanted a conversation with another European from outside.

They had the lovely and incredible knack of landing and talking to folks under most weather conditions.

"Why incredible?" Did I hear you ask? You try landing a Cessna from about two-thousand feet when you can't see the ground – I'm fair dinkum.

Remember it's the wet season and we have complete cloud cover on most days.

These RAAF trained pilots would know when they were over the village, even though they can't see the ground.

"Impossible!" Did I hear you say? This is what they would do. While above the clouds, they would circle and drop about fifty-feet each time. We could hear them circling overhead and there was never a lack of welcoming committee, let me tell you.

After a circuit or two, they could still be in cloud with zero visibility. They knew how high they were by their instruments. They would continue this circling and dropping fifty-feet per circuit until one of two things happened.

I am not having you on.

If they found a hole in the cloud – they'd bank sharply and put the thing on the ground. By jingoes, you need nerves of steel.

If they struck the upper branches of a tree, they would cancel the landing and fly off.

How do I know all this, you ask breathlessly. It was my job to check the rear wheel and, if there were any small branches stuck in the wheel assembly, I had to pull them out. And I did this on several occasions.

There was a functioning bauxite mine, inland from the mission, and the people who were running the mine decided they needed a more abundant supply of water for the processing side of things. Bauxite, by the way, is the source material for the creation of aluminium, and apparently Australia has plenty of it to sell overseas.

A Melbourne drilling concern won the contract to come up to the Cape York Peninsular to find and drill for the wished for water.

I well remember, even though it is over fifty years ago, the huge amount of equipment that came through the mission from the Archer River landing area, only to disappear eastwards towards the mine.

On one occasion, I met one of the drillers and asked, "How far down can you guys go with this little lot?"

"We can go down five miles, no problems."

A few months later, all the machinery and drill rods started to come back towards the landing area. I asked one of the drillers, "How far down did youse go?"

"Five miles and we found nothing!"

The head contactor asked to see the mission boss. He asked, "Is there anything we can do as payment for using your landing area?"

"As a matter of fact, yes," replied our mission superintendent. "We would like a little more water ourselves for the people."

"You'll have to tell us where to drill, but no problem."

Youse, yous or y'all
single or plural word for referring to another person or persons.
"Can yous all help me with this?"

Some Aboriginal elders were asked where would be the best place to drill for usable water for the mission houses.

They indicated a spot close by our night soil station and almost two-hundred feet from the edge of the river.

I'm not sure what went on but I do know that the drillers were quite upset because water was struck at nineteen feet. That's not even one drill rod length.

At the mine no water was found, but at the village, usable water was a mere nineteen feet down.

Those elders sure know what's what when it comes to understanding the land.

You can imagine the dummy spit the head contractor had.

Dummy Spit
a childish display of
exasperation or bad temper.

Part 2: People

**Where the people hurt and cry
You know, well, that's just wrong.**

Spinster from Paradise Station

In the vast New England of northern NSW, a horse drawn coach had picked up some passengers. One of these was a spinster who worked on one of the many cattle stations.

Driving a coach drawn by a team of horses was quite a skilled job, especially in poor light as they only had carbide lamps. Headlamps were yet to be thought of, let alone invented.

The coach driver was heading to town as per normal. He was somewhere near Elsmore when he got a shock at the words: "Bail Up! Bail Up!!"

A notorious local bushranger had come in the early evening gloom and startled the driver.

The passengers quickly stretched the leather flap that passed as a door on these early coaches, trying to hide from view.

"I want all the men's wallets and I want to kiss all the women!"

The driver, having recovered his composure, began to argue, but hesitated when he saw the gun.

"You can have our wallets, but leave the women alone!"

The argy-bargy continued until the spinster leaned through the leather flap and began to hit the driver on his arm with her umbrella.

"Driver, just who's holding us up – you or the bushranger?"

I think she wanted a kiss from the bushranger. Madness? Oh yeah.

Maybe she wanted to get close enough to see who he really was!?!?

That would go over well at the next CWA meeting.

Quick Quiz:
What does CWA stand for?

A. Chicks With Altitude - *we are the highest branch in Australia*
B. Chicks With Attitude
C. Chin Waggers' Association
D. Country Women's Association

Argy bargy
argumentative talk, wrangling

Christmas at Cheesering

Guy Fawkes night approaches. I'm seven years old and determined not to miss out on a bonfire for November 5th 1953.

As any over enthusiastic seven-year-old would do, I started in on building a bonfire. Why wait for big brother?

Pieces of timber were few and far between, but why leave all these metal bits lying around? Anything that was easily lifted, was – and was deposited on my bonfire. Even some galvanised iron shards were dutifully deposited on my bonfire pile, ready for the special bonfire night.

Have you ever had a little brother who enjoyed chasing you around? Have you ever had a little brother that you could easily outrun. I did.

During one of these hide-and-seek chases, I decided to use my bonfire to hide behind. Mistake. During the ensuing run around, I tripped and fell onto some galvanised iron; its sharp edges tearing some respectable wounds in my right knee. I can see the scars to this day.

I assume you've heard about blood poisoning, right? I found out about it because of that incident.

Long story short, I found myself in Liskeard hospital for Christmas.

I have the lasting impression that I was not expected to survive because my bed, the cupboard beside the bed and the windowsill were covered with Christmas presents and cards.

My young eyes could hardly take in this feast of presents. And to this day, I regard this as my best Christmas ever.

When I finally arrived home, I was allowed to open my presents.

Dad said to me, "Here's one from my sister, your Aunt Nancy."

She sent me three shirts. I seldom received anything new when I was young.

In the Upton Cross Primary School, they encouraged us to keep a diary. Obviously, I made a mention of receiving this very welcome, but unusual, gift.

End of story.

Not really.

Many years later, when we had emigrated into the Aussie bush near a place called

Elsmore, my mother unearthed this school diary and had quite a chuckle.

She was still smiling when she brought the little book to show me.

You see, I had misspelled the word, 'shirts'. I had left out the letter 'r'.

I was only seven or eight years old – give me a break!

Aussie
a person from Australia (not often used to describe the country of Australia)

What do you eat when you're lost in the desert and hungry?

You eat the 'sand which' is there.

The Butcher's Name was Mr Push

I started my working life as a farmhand working for the Woodland family. They always invited me in for lunch.

On this particular day we had a visiting couple from Victoria who had wanted to go and explore this northern NSW town.

In the process, they were asked by their hostess to purchase the meat Mrs Woodland would need for the evening meals.

Apparently, the quality of the meat was superb and, over lunch, the visitors were being questioned as to which butcher they had purchased the meat.

"Was it near the fire station?"

"No, I don't think so."

"Could you see the town hall from the butchers?"

"No, well I didn't."

"Was there a pedestrian crossing near his front door?"

Farmhand
a person hired to work on a farm

The questioning subsided into an impolite silence.

It just so happened that the Woodland's eleven-year-old daughter had accompanied the Victorian visitors on their trip to town.

She had remained quiet through the interrogation. Until now.

"I know the butcher's name!"

Delighted, the family and visitors all begged her to share the detail that none of them could recall.

"It's Mr Push. It was written on his door."

We tried awfully hard not to laugh and embarrass her.

Long Distance Vision

An old farmer of many decades was having trouble with his eyesight. On his next trip into town he found himself in the optometrist's waiting room.

Not having an appointment, the farmer had to wait, but the optometrist eventually got to see him.

He asked him to come in and sit in the special chair. You know the drill, they swing a special jigger in front of your face and ask when they turn the wheel thingy, "Is that better? Is that worse?"

He performed all the normal tests and eventually swung the apparatus away and

---

Jigger
someone who jigs (a type of dance). It can also refer to just about any mechanical equipment that jolts.

Thingy
an object or event which the speaker can not (or does not wish to) identify.

announced, "I'm sorry sir, but you have normal wear and tear for your age; nothing can be done."

"Well I'm not happy," said the farmer as he left the office.

A month or so went by and the old guy showed up again.

"What can I do for you today, sir?"

"I still can't see far enough," the old fellow replied.

"Look, come in and sit down, I'll run the test again, I might have missed something."

Same result. "Sir, it's just wear and tear appropriate for your age."

"Well I'm not happy."

"Sir, you'll just have to accept it, there's very little that I can do."

"Well I'm not happy!"

The optometrist realise he has problem, and then had 'a lightbulb moment'.

"Sir, come outside with me."

They both stood on the footpath, and the optometrist pointed up to the sky and asked, "What do you call the big bright thing up there?"

The old guy, shielding his face replied, "Why that's the sun".

"Well sir," responded our friend the optometrist, "Just how far do you want to see?"

Dad's Old Station Wagon

Somewhere out near Ghoolendaadi, a farmer and his wife had three daughters. Two were in high school and one in her final year at primary school.

They were a hardworking, peaceful family with only one topic that brought conflict ... Dad's old car (bomb).

He'd had this old station wagon since his younger days, and he knew how to keep it going whenever it complained.

But he knew, and the girls knew, that age was catching up with the old thing.

This crisis got to the point where the girls hated going into town with dad. At least mum knew where to park inconspicuously so no-one would see them.

Farmer dad could see their point of view and understood that his daughters' social standing was besmirched by the appearance of the old girl in the streets of Gunnedah.

Dad could see its days were numbered.

"Darling," his wife began. "I tore my dress getting out of it last week!"

Dad knew he had a problem. He'd been chewing this over for weeks now.

At the family dinner table on Friday night, Dad made an announcement, "Girls, I've made up me mind. As soon as I get two good harvests in a row, I'll buy us a new car that'll you'll like."

WELL.

Dad wasn't prepared for the avalanche of squeals, hugs and kisses from his daughters. Mother just smiled and looked on lovingly.

The day came when the second good wheat cheque hit the bank. True to his word, dad went into town and to the car dealership.

They agreed on a trade valuation, placed an order and paid a deposit on a sedan he knew his four girls would like. Another high point in dad's popularity.

Eventually, the long-awaited phone call came, "Your new car is ready for you".

Can you imagine the glee this news brought to the household.

As he was paying for the new one and handing over the ignition key for the old bomb, there was a pause in proceedings.

"Look," he said. "I do have a second key for the old girl, but I looked this morning and wasn't able to find it."

"That's fine," the salesman replied. "Just bring it in when you find it."

As you can imagine, when dad pulled up at the front gate, the girls launched themselves at him.

They could scarcely take it in. "A new car for us."

And, once again, there were kisses and cuddles aplenty for dad.

The girls' social standing received a huge boost.

Now, it just so happened that he did find the second key for the old wagon and, to be sure he wouldn't forget it on his next trip into town, he put it onto the ignition key ring which had the new car keys on it.

Some time later, dad had to go into the Seed and Grain merchant and order what was needed for the next year's sowing.

Now look here chaps, there's a lot to be thought through when planning for next year's crop.

So, what with the mental calculations, and the nagging thought of the new tyre that will be needed before long on the seed combine, dad had a lot on his mind (no farmer wants to be held up with the sowing when there's rain on the way).

So, as he left the Seed and Grain and headed for the car, dad's mind was somewhere else.

As can only happen in a country town, the proud new owner of dad's old bomb had parked beside the new car that the girls were so proud of.

With his mind on the upcoming harvest, dad hopped into the old thing (oblivious to the fact that this car now belonged to someone else) and, as was his usual habit, headed home.

He pulled up in front of the house thinking, 'Ah, I forgot about crop insurance'.

The girls, on seeing the 'old bomb' outside again, launched themselves at dad - but not with cuddles and kisses this time.

"You silly old fool, what happened to our new car?"

"Dad, did you have an accident? What happened to our new car?"

Dad was in mid-thought. Then grasping the import of their disgust, turned and glanced at the new car.

NO! It was the old one!!

"Holy cow, I forgot to hand in the old ignition key."

Whatever parcels and groceries he had bought, were thrown out onto the ground (with colourful language). He quickly hopped in, drove it back to town, and parked it beside his new car. Then he drove home, let's say ... very quickly.

Instant forgiveness and peace once again settled on the household.

Fokker Friendship

Somewhere in NSW, an east-west airline, Fokker Friendship, was flying when to the pilot's dismay, the aircraft stopped responding to the flight controls.

A frantic call was made to the Tamworth control tower, but no-one was available who knew about this aircraft. Can you imagine the panic when the pilot realised that the aircraft was nose-down?

"This thing is going in if we can't figure this out." And yes, it was a commercial flight with passengers on board.

Realising the danger, the Tamworth controllers banned all unnecessary voice traffic.

During the frenetic activity in the Fokker's cockpit, the pilot heard a radio voice of someone trying to talk to him.

Tamworth kept threatening the caller to stay out of this emergency.

The caller persisted, "I KNOW WHAT'S WRONG, this happened to me!"

Tamworth tower said, "Okay. Go ahead."

To the pilot of the nose-down craft, he asked, "Have you moved the pilot's seat at all?"

"Well, yes I have!"

"Right. Hop back into the seat and move it forward – all the way."

"Okay."

"Now, hop out and look behind the seat. You'll find a flight manual that's fallen off the shelf above your head; remove it and you should have normal control of the aircraft."

He waited on the line for a response.

"Yes I do!" The bloke was nearly in tears. "How can I thank you??"

"No probs mate, you were lucky enough to be close to where I am and I could hear the panic. Stay safe my friend."

The Fokker Friendship crew were beyond words.

You see, when a metal covered object falls off the high shelf, and the pilot moves his seat back for leg room, the object can jam up against the master hydraulic control valve which eliminates all hydraulic control valve function.

Apart from this, I always enjoyed flying in the Fokker Friendship.

Probs
A lazy way of saying
"probably" or "problem"

He Saved Lives by Doing Nothing

Barry was one of those rare blokes who did everything well. He was largely self-taught but, by jingoes, he was very smart and extemely intelligent. The Aussie bush hides plenty of these folks who never got past primary school but who are masters at nutting out a problem.

Apparently, General Montgomery had intentionally picked the Aussies to lead an advance because of our native ability to do just that: find a solution to a problem or delay.

Back in the early 1970s Barry was asked to drive a large earthmoving dump truck from Port Hedland to Inverell northern NSW. We called these monsters 'belly-dumps.'

They consisted of a large tractor-like prime mover with smaller wheels on the front which were the steerable wheels for turning corners.

The large rear wheels stood taller than most men and were driven by a blown

V12 diesel engine which could produce something like 500HP.

From behind the driver's cabin, a large gooseneck formed the front of the huge, long rectangular bowl which was supported by the two rear wheels.

When in use, this large cavernous bowl would be filled with, let's say, dirt – and then transported away from the construction site and emptied by the two long doors forming the bottom of the bowl.

They could easily carry thirty-five tonnes of dirt just to deposit it somewhere else. This must be done on the move, otherwise the machine would be stuck on a mountain of dirt when the doors opened.

So, by now you can easily imagine that such vehicles were not intended for working on our public roads. But our hero, Barry, had been asked to drive this large dump truck halfway around Australia – on public roads.

This was a task that only Barry would be good at, and he was eager to start.

Rear-vision mirrors were fitted, stop lights, reversing lights, witches hats – you name it. Even drinking water was supplied.

His employer had every confidence in him – and rightly so.

So off he set, heading south from Port Hedland. I'm guessing he plotted his course from motel to motel where he would rest each night – knowing this is going to take many days.

Okay, so these vehicles were just too big for most roads, and dangerous on smaller roads but, by the living Harry's, you couldn't ask for a better operator than Barry. With Barry at the wheel he kept things as safe as they could be.

This was soon to be proven correct.

All was going well until he entered NSW. He had to use the New England Highway, not his first choice, and there you have it!

Fasten your seatbelt dear reader, you're about to experience a total mental and bodily foul up. But he wasn't even laughing when he told me this, back in the early 70's.

Somewhere near Aberdeen, he was approaching a single-lane bridge. These were the true dinosaurs of the era of early convict-built road bridges. They were narrow and dangerous even for normal use; and on this day, Barry's driving a vehicle that's nearly as wide as the bridge itself.

Fortunately, he was heading north, and the southbound traffic could see him coming from quite some distance. He, being northbound, already had a substantial train of cars behind him. He still had to stop and allow some stragglers to clear the bridge before he could proceed.

So, here we are – lined up for the bridge – the other southbound vehicle drivers had stopped short of the bridge, either out of caution or fright at the sight of this huge belly-dump coming the other way.

You haven't had that mental and body foul up yet; check your seatbelt, please.

I have warned you!

He would have been about fifty metres from actually entering the bridge when, to his gut-wrenching horror, he saw one of those early British built, low sports cars on the outside of him and rapidly overtaking him. I think we used to call them MGB's.

Barry's front wheels found the edge of the bridge pavement. The overtaking driver was commited to the overtake. There was absolutely *no room* even for half the carwidth of the MGB.

The whole tractor part of the belly dump is now on the bridge.

CRIKEY!!

Barry looked at the rear-vision mirror again.

Nothing. There wasn't a thing in sight! Where the B - - - - y hell are they?

Oh NO! Have they been run over by these rear wheels?

Have they speared off into the river?

Are they stuck under this truck?

Have they - - - - - - ????

Should I slow down?

Should I knock her into neutral and give them a chance to survive !?!?

77

By sheer instinct, he froze and didn't move anything; no lever; nothing. Even his foot on the accelerator didn't move.

But they were gone. God knows where.

Have I just killed some people?

By now he had just come off the far end of the bridge, still frozen at the thought of there being people back there in a crushed little sports car.

He was about to check his mirrors again when this little undamaged sports car slid out from under this monstrosity and shot off into the distance at half the speed of light.

Okay dear reader, you can relax now, but there'll be no laughter thank you very much.

We haven't finished with Barry yet.

He became the head man in the lube bay on Copeton Dam. This meant he was responsible for all the engine oil changes and the general maintenance of the fleet of rear-dumps and belly-dump trucks, not to mention light trucks and cars.

One consistent complaint about one particular dump truck was that the engine always overheated. Not to the destruction stage, but enough to worry whoever was driving it.

Lube bay
the area in a mechanical workshop used for lubricating vehicles.

As a driver, you are responsible for the vehicle; and every driver hated driving this thing.

You will find this hard to believe.

Barry decided that he would personally do this engine oil change. Normally, he was too involved in the oversight of things to actually get his hands oily as is the case when you're draining and refilling about twenty litres of oil.

So, we have the oil drained and the sump plug back in, we have the new oil filter primed and mounted.

Okay, let's fill this thing up with twenty litres of fresh oil.

Wait a minute. "I've just put twenty litres in and it's only just half full. All the other trucks of this type take exactly twenty litres."

What the??

Another ten litres and the dipstick reads full. What's going on here?

They don't make dipsticks like they used to in the old days!

"There's something wrong with this dipstick" he said to himself.

He took a dipstick from another truck the same as this one, dipped it in and read it. The damn thinging is now way over full!

The light was beginning to shine in his head. By placing both dipsticks side by side, he could see that the one that belonged to the truck he was working on was only two-thirds the length of the one from the other truck.

Ahha.

This damn truck has been running around for years with the wrong length dipstick.

So this is why it's always running hot – the oil is being constantly whipped up by the immersed crankshaft.

It's a wonder it lasted this long.

He adjusted the oil level using the correct dipstick and ... guess what?

Well I'm not a genius, are you???

Just wanting to help

A group of young adults from overseas were taking a joy ride over barren parts of Australia in a small plane.

Two chaps up the front were discussing the problems down there when, unexpectantly, one of them opened the window beside him. He fished something out of his pocket and was about to throw it out when the chap beside him grabbed his arm and asked, "What are you doing?"

"I'm going to throw out this $10 note and make someone happy."

"Let's think about this. Why not throw out two $5 notes and make two people happy."

"Okay." So he found two $5 notes in his wallet and reached for the window again. At the point of releasing them he was stopped again. "What's the matter now?"

"Why don't we throw ten $1 notes and make ten people happy?"

A voice came from the back of the cabin, "Why don't you throw yourselves out and make everybody happy?"

Charlie's Toilet

Farming in Aussie is very labour intensive.

Charlie and his wife were no exceptions to this and were constantly on the go.

But, to have some relaxation on the land is also important and they knew this – which is why they decided to build a tennis court in the disused corner of the home paddock.

It was more than big enough for the project, so work was started.

First, they decided to kill off the tough weeds and small trees growing right where they don't need them.

Tennis court gravel was delivered, and it was all-hands-on-deck to spread this stuff out and then to level it off.

When I say 'all hands', I mean exactly that. Men, women, teenagers and young folks all had a job to do.

Very little on the land gets accomplished without the female input. She's working in the kitchen preparing lunches, she has clothing ready, and keeps everything going. She's on the shovel, repairing damaged fingers or hands, running little Gertrude to the doctors, or meeting the school bus. She keep the workers on the job.

As you may have guessed, the project required all the local families to participate.

"After all, this is our tennis court not just so-and-so's."

With work progressing, every second or third weekend (the farming can't be stopped totally) Charlie was concerned that they didn't have a toilet should mother nature call.

So, he took it upon himself to build a reasonable size garden shed to encompass an actual toilet, and a small storage shed to keep the net and other important equipment in.

After the first weekend on the shed, he had the floor and one wall erected. When onlookers saw this they asked, "How long before we can use the facilities, Charlie?"

He had a sense of humour that wasn't always on show, and a mischievous twinkle in his eye if you were quick enough to observe it.

"It should be ready this week. I'll make up a sign, 'Women 'round the front and men 'round the back!'"

That joke was not appreciated by anybody. (It makes you wonder why I wrote it in the story now, doesn't it.)

Needless to say the shed, toilet and tennis court were completed and enjoyed by all.

Freight On!

A farmer friend was quite used to working all day in the hot Aussie sun.

I was ploughing a paddock one day and watched as he loaded bags of wheat onto the tray of his truck. We're going back a few decades here. A bag loader is a hydraulic arm that's activated as you drop a bag of wheat onto it's lift platform.

This triggers the hydraulics to lift the bag up in a pivot motion until it reaches it's maximum travel where the bag can be off-loaded and placed in an orderly row of bags already on the truck. When the weight of the bag disappears, the hydraulics reverse and the lift arm comes back down to the ground ready for the next lift.

The reason for this lengthy explanation is this: as you can imagine, the tray is over a metre above the ground – hence the need for a bag loader – but it doesn't come with a step ladder for the farmer to get up onto the truck to receive the next bag.

This is where it got interesting for me to watch as I drove the tractor for ploughing. You see, what my friend would do – time and time again, and for hours, mind you – is go up with the bag.

He would athletically step onto the lift arm close to the pivot, and allow himself to be lifted up with the bag; then quickly step off the lift arm just in time to receive the bag of wheat onto his shoulder and drop it into place with all the other bags.

He would then ride it down again, drag another bag to the loader and drop it onto the lift platform.

He would do this repeatedly until there was no more room on the truck, and he would drive off to the wheat shed.

But, growing wheat wasn't his only skill. I can assure you that all successful farmers have more than one skill. This farmer's other skill was as a sheep farmer.

If you wish to farm sheep for their wool, then sooner or later you'll need to employ shearers to take the wool off.

There are many classes of wool, and usually the woolgrower himself separates the good wool from the stained or damaged wool, which is kept separate.

In this particular year, he found himself with an abundant amount of wool which he industriously set about baling and preparing for the local buyer to collect and send off to the large wool mills overseas, usually Japan.

I'm only guessing how many bales he would have ready for shipment, let's say around fifteen. They were duly picked up and would eventually arrive by ship at the wool mill.

I haven't the faintest idea how the local wool buyer would arrive at a value, but both parties would agree on a price, and that was what the farmer was paid. No problem.

But this year there was a problem. A big problem.

What I haven't told you is that my friend would take a large cordial bottle of cold tea every day, to sustain him until the next mealtime.

In this particular year, he misplaced the bottle of cold tea during the baling. He searched high and low, but gave up hope of finding it.

Many months later, the local wool buyer rang my friend and enquired, 'Are you missing a cordial bottle by any chance?'

"Yes, why?"

"It was found in one of your bales and they have sent it back – Freight On!"

"Apparently they were quite upset, and the freight is *very* expensive."

Freight on
the cost of shipping / postage / transportation has been added
---

A Beekeeper in Trouble

This story involves a worried beekeeper and starts somewhere on the New England Highway.

Our poor friendly beekeeper had been suffering from a toothache for a few days.

His bees were onto a good flow from a patch of Yellow Box trees, and he knew the supers (the top box reserved for storing nectar) would be nearing full by now.

"Toothache be damned," said he, and he loaded the truck with enough empty supers for the job and headed up the New England Highway.

The toothache wasn't getting any better, *But I can't just leave the supers on the hives when they're full.*

Perseverance got him there and, sure enough, the full supers were ready to be replaced with empty ones.

This would usually take two to three hours, but today this was taking much longer. Trying not to hang onto his sore jaw, he managed to get the job done.

Replaced and loaded, he now had to rope the piled supers in place ready for the trip home.

The toothache was not getting any better!

I'm guessing the ropes he used to secure the load of precious rewarding nectar were not as tight or arranged neatly as he would usually do. It's hard to do a two-handed job with only one useable hand.

As his wheels hit the highway, the pain became nearly unbearable. *I have to do something,* he promised himself. But he knew that dentists were hard to find in the bush.

As he was getting closer to Tenterfield, he decided to stop at the service station and ask for the nearest dentist.

He half-limped into the service counter. The bloke behind the counter took the situation in at a glance.

"Crikey mate, you need a dentist!"

The beekeeper tried to ask, but the bloke answered, "There's none around here. You're best bet is to get to Glen Innes".

As he turned to leave, a local came in and saw the distress of the beekeeper.

"Tell you what, head south and take the turnoff to …" (I'm not sure what road he mentioned) "… you'll know it by the huge apple tree on your left as you turn. There's a retired dentist about two k's down that road. He lives in the only two-story house hereabouts."

What choice did our pain-ravaged friend have? He headed south and found the turnoff with the humungous apple tree.

Apple tree honey is not good honey and we used to avoid it.

He found the large house.

There was a bloke working in his garden as the world-of-pain beekeeper approached.

"Are you the retired dentist?"

"I am."

"Could you please help me?" He was trying to speak English, but it can be difficult when your facial muscles don't work the way they should.

Again, his condition was obvious to all. The dentist stood up and ushered him into his front room. He gave a quick examination and left the room.

What now? thought the beekeeper.

Afterwards, when recounting his painful adventure to his family, "I suddenly felt a pair of pliers in my mouth, and out she came. I've never felt such relief in my life."

K's or k's
kilometres, the Australian version of measuring distance.

He was given some form of medication, then thanked the fellow profusely and headed home, in much better shape than when he'd started.

I'm guessing the retired dentist would have refused any offer of payment. The obvious relief on his face would have been enough.

The Worm Song

This song has been around for a long time, but please don't teach it to your children!!!

Nobody loves me
Everybody hates me
Think I'll go and eat worms.
Long skinny ones
Big Fat juicy ones
See how they wriggle and squirm

Bite their heads off
Suck their blood out
Throw their skins away
Nobody knows how well we live
On worms three times a day.

Can we help you?

Living in the country usually means not buying your entertainment – you have to invent it.

One full moon night, little brother and I decided to go for a long walk. We had noticed before on equally bright moonlit nights two small but distinct hills on the horizon about ten kilometres away.

So tonight, we agreed to set off after dinner and see if we could reach those hills. The fact that we were walking over private farmland didn't seem to worry us.

After about an hour of walking and climbing several fences, we found ourselves striding down a gently sloping paddock which had obviously been recently harvested. The short stubble made it easy for us to increase our walking speed.

Quite unexpectantly and suddenly we both, in perfect unison, swayed backwards and halted. We hadn't seen a single wire fence traversing the paddock from our left to our right with a steel post about every thirty or forty metres apart.

We couldn't believe how we both unconsciously became aware of the wire about chest high and reacted before actually seeing the wire and recognising it for what it was.

Bear with me please reader; this true story has a sad ending.

We gathered our wits and pressed on through more fences and gullies.

After about two hours since leaving home, we decided that those two hills weren't getting closer. Without debate, we decided to turn around and head for home.

PROBLEM.

We couldn't recognize anything in the distance that looked like home. It was right here that I learnt a very valuable lesson. When you're walking in the Aussie bush, day or night, you must look behind you every two hundred or three hundred metres. This way you know what the way back looks like.

We just walked, hoping we were going 180 degrees to the way we had come. Now – that's a recipe for disaster – isn't it?

We had walked the best part of an hour (yes, another hour) when we crested a hill and could see light in the distance. Relief flooded both of us.

"Ahha," said I. "Mum and Dad have put a Tilley lamp out for us."

This proved to be very wrong, and I regret what happened next.

From here we had only gone a short distance towards home when we both heard something at the same time.

"Sounds like an engine running," said Richard. We continued towards home, and the engine noise got slightly louder.

"Who would be running an engine out here when it's well after midnight?"

We continued on, and through another fence, then we could suddenly see the source of the noise.

Off to our right, there was a stationary tractor, engine idling over, rear spot lights on, but nobody in sight. Truly, this is a strange and weird scenario in the middle of the night – especially with no living thing in sight.

We were still staring at this strange sight when we saw movement under the plough which was hitched to the tractor.

"CRIKEY! It's a bloke and he's been run over by the plough. Struth!"

Struth
an exclamation expressing
surprise or verification.

Our young legs got us over the newly ploughed soil and up to the plough. There's a bloke under the plough alright and he's writhing in agony.

Bear in mind, it's the wee hours of the morning and as far as the accident victim is concerned, there's no-one else on the planet! If he can't extricate himself, he'll be there till day break at least.

We quickly asked, "Can we help you?"

To say the bloke became instantly terrified is a galactic understatement.

What hurts me to this day is the sickening sound of his bald head ricocheting off the steel frame of the plough. Three times his head impacted that solid steel frame as he tried to make sense of the dark night speaking to him.

We helped to drag him out from under the plough. It's here we realised he hadn't been run over at all. We also realised he only had one arm.

For quite some time, he still wasn't able to talk to us. Then we realised what the true situation was. He slowly recovered the power of speech and explained what we saw.

Here was a farmer with only one arm and his plough had become clogged with a massive tangle of castor oil weed. We apologised for giving him such a fright.

"Can we clear the plough for you?"

He nodded in agreement, "Ye-yes, okay, thanks."

We made short work of the blockage as we used the short pole he had with a hook on one end.

"Wha – what are you two doing out here in the night?"

"We've been for a walk but we're not sure where home is."

"Where do you live?"

"At Avondale on Charlie Woodland's place."

"Oh," said he. "Head in that direction." He was pointing in a direction which was on a complete tangent to the line we had walked in on.

We said our goodbyes. He climbed onto the tractor and continued ploughing, and we headed home.

We discovered no lamp for us – no problem. We opened our seldom-locked front door, climbed into bed and passed out.

Japanese Tourists

A long-time friend of mine was approached by the manager of a local tourist company to welcome into their home a middle-aged Japanese couple. The couple were visiting Australia for the first time and having trouble dealing with the Aussie life.

They were getting disoriented, causing them to get lost in town; and they were alarmed at the strange life we have here.

Farmer Isaac and his wife were only too happy to accommodate them for one or two days as it was felt that the couple would find life much easier in a normal house with a family. Surely this would be more to their liking.

The theory was good, but Isaac's home was far from normal and fairly large even by Aussie standards.

The guests were welcomed in with typical Aussie generosity and shown their rooms. Our hosts didn't twig that there was a problem. Shortly after, the visitors seemed unsettled and came back to the lounge room looking for something.

Somehow it became obvious they were in need of a toilet.

"Oh, go down that hallway and it's the first door past the laundry."

Mrs Isaac quickly led them down to the correct door, and their relief was plain to see.

Fifteen or twenty minutes went by and they didn't come back! Mrs Isaac went looking for them, they weren't in the toilet!

But she could hear them talking! Were they in the sewing room? No. Oh, they're in the double garage! She led them back to the lounge room.

"Are you okay?"

Their English was good enough to make themselves understood: "Ah, all too big!"

Mrs Isaac began to prepare a simple evening meal and their female guest tried to help. She found the kitchen bench too high and was confused by all the drawers and such a large refrigerator.

The meal was consumed with polite and simple word conversation.

Bedtime arrived and our guests headed to their rooms.

Perhaps mother nature required attention, I'm not sure now, but navigating to the toilets and back to their rooms didn't go smoothly.

Mrs Isaac came to the rescue again and the house settled for the night.

Somewhere along the way our farmer host promised them a tour of the farm tomorrow – this would prove to be a mistake.

All successful farmers are early risers and Isaac was certainly one of these. The visitors too were early risers and joined their hosts for breakfast.

Said Isaac, our thoughtful Aussie host, "After breakfast, I'll take you for a brief tour of our property."

"Ah, yes please."

They climbed into Isaac's car, and he headed out the gate, did a left turn, and drove towards the small road that ran through the back of his place.

"Who lives in that house?"

"Oh that's just an old house we haven't used for years."

"You have a house on your farm and no-one lives there?" Mrs Visitor asked incredulously.

Replies our host, "Oh we will tidy it up one day and use it."

The visitors were a little unsettled at this, but didn't comment further.

The car continued until Isaac crossed the road and, after opening a gate, he continued into the rear part of his property.

"Where are we now?" Asked our visitors.

"This is just the beginning of our back paddocks."

To their understanding it seemed as if they were entering another country!

Isaac could see that his visitors were not enjoying things as he had hoped.

"We'll head back now for morning tea."

A smile appeared, telling Isaac that he had struck the right note at last.

Over morning tea, Mrs Isaac suggested that they both might like to go for a little drive themselves.

"But no licence!"

"You don't need one provided you stay on our property."

"Okay, thank you."

I have no idea what happened next, but Mr Visitor was obviously quite competent at driving, so off they set for a little joy ride of their own. Perhaps they thought this would be the highlight of their farm stay downunder.

They were enjoying themselves at last in this wonderful, but hard to understand, country.

Downunder
relating to Australia or New Zealand

Perhaps there are no contour banks back home or they've just never seen one before.

Let's see what happens when you drive over one.

MISTAKE.

The vehicle made it halfway over the contour bank and got stuck! Bogged on a contour bank. It does happen.

The vehicle had no traction. They were going nowhere!!

They were about two kilometres from the house. It's about here that they noticed they should be back in town to catch the plane back to Sydney very soon!!

I'm told they ran back to the house.

"We'll miss the plane!!"

I'm not sure what condition they were in, but they were bundled into the car – luggage and all – and made it just in time.

Australia isn't for everyone.

Part 3: Truth

**When our words get all mixed up
We can't tell dark from light.**

The Bendemeer Bridge

Way back in the days of horse and bullock teams, the roads in northern NSW were still being made or upgraded to suit wheeled wagons.

Travelling south on the New England Highway, as you approach a town called Bendemeer, you have to cross the Macdonald River Bridge. The problem here is the road takes a ninety-degree bend to get onto the bridge.

No problem for today's vehicles, but the bullock teams could be fifty to eighty metres long, plus the length of the dray.

The often-repeated joke by the team drivers was,

"We'll have to ... bend 'em here fellas."

Say it quickly and you'll figure out what the bullockies were laughing at.

Bullockies
those who drive the bullocks
(a castrated male of a bovine animal)

Elizabeth, Lizzy, Betty and Bess

Went over the water to find a birds nest

They found a birds nest with 4 eggs in it

Took one each and left 3 in it.

How come?

Dad's New Tractor

The eleven-year-old son was sure he would follow in dad's footsteps as soon as he finished school.

Dad would often remark how useful he was around the farm even at this young age. He was especially good at working with farm animals. He seemed to really understand them.

At breakfast on this particular morning, the young lad overheard Dad arranging for the local farm machinery dealer to come out and service the brand new tractor.

Any new tractor must have it's first service done by the dealer that sold it.

Junior fell silent and wouldn't respond when asked if he wanted more of something to eat. This was very unusual and both parents tried to get him to answer them – but he was completely silent.

He was like this for the whole day.

At dinner, his parents tried again, "Darling whatever is the matter? Your father and I are quite concerned. Has something upset you?"

At breakfast next day he finally spoke.

The young lad was very proud of Dad's new tractor – he'd even been allowed to drive it.

"I'm upset about them coming out here to service Dad's new tractor."

"But son they have too, it's part of the contract."

"Well Dad, I seen what happened when old Murphy's bull serviced our Betsy the cow. I'm not having that happen to our new tractor."

I'm Nott, the Rabbit Inspector

Mum and Dad were well-respected farmers, somewhere out near Bingara, in the days of the rabbit plague.

Farmer Dad had been asked by the family (for some time now) for a short holiday on the coast.

"But who can we get to look after the place while we're away?"

Dad remembered his cousin, a townie, who agreed to fill in while the family had a long-overdue break.

While they were loading up the Land Cruiser, Dad turned to his cousin.

"Look, if anybody comes to look at the place, don't let them past the front gate. I've been able to clean out the pests in the front of the farm – you know, the ones that can be seen from the road – but haven't had time to clean up the back paddocks yet."

Dad forgot to mention to his cousin that the name of the rabbit inspector was Allan Nott.

Loading completed, off they set with a wave goodbye.

Cousin retreated to the house. During the week, he heard a vehicle pull up out the front. By the time he got to the front door, the stranger was already halfway to the door. The friendly visitor held out his hand and introduced himself.

"G'day, I'm Nott, the rabbit inspector."

Cousin was so relieved at the news that he shook his hand and then wiped his brow.

"Thank God," said he, "There's thousands of them out the back".

Stuck - Out Here!

Our travelling salesman friend was doing the dreaded out-west-run selling his wares when, on hearing of rain, discovered that all the roads east were cut.

Stooging around, he eventually found a motel, slightly run down, but at least they had a vacancy.

He booked himself in and settled down in his room to await events.

The next morning he went down to the breakfast room and asked for the menu. Ordering bacon and eggs, he noticed that there was something peculiar about the menu; but couldn't quite figure out what.

The bacon and eggs were duly delivered and he waged in. Suddenly, it occurred to him what was strange about the menu; all the meat dishes were pig meat.

"Yeah, fair enough," he said to himself. "I suppose I'm a fair way out here and, if the trucks can't get through, well, they have to live off the land. Yep, that's fair enough."

Having consumed the first course, he ordered some tea and toast.

With a bite of toast in his mouth he took a swig of the tea. "Ew!" he spluttered, nearly spitting it all out on the floor.

Turning to the young waitress he demanded, "What's wrong with the tea?"

"Oh, don't worry about that sir," she replied. "That's just the bore water."

"STRUTH!" he exclaimed. "You don't waste much do you?"

What do you get
when you pour
boiling water
down a rabbit hole?

Hot cross bunnies!

I Don't Understand

Let's get one thing straight: horses and I don't agree. I've never been able to get along with horses; it's like an undeclared war between us.

But long ago, horses were the main form of transport.

Unlike today, not many people had a car. I'm thinking late 19th century and early 20th century.

I'm glad I wasn't around back then.

I was told once to regard a horse as a human – just in a different form. Make of that what you will.

One of my early employers came from a Germanic family who were part of a mass exodus of Europeans who came to Australia in the early years of our country. They disembarked in South Australia and moved eastward and then up into New South Wales and some got as far as Queensland.

You will still find their names in the local phone book today.

My early employer's family settled in the Geelong part of Victoria and, being an industrious people, soon had a coach building (I think) workshop up and running where repairs and maintenace on horse buggies and coaches were carried out.

The senior men had learned their skills back in Europe and proved to be a real asset to the local communities. Major repairs were done with admirable skill and safety.

One day my boss' father was asked to quote a repair on the main structural member of a horse buggy.

This large timber board spanned the width of the buggy and was also the rear seat support for the driver and passengers. I'm guessing that the structural member would be somewhere between fifty and sixty millimeters thick.

If this large board is allowed to wear or become damaged, the safety of the whole buggy (and any passengers) was compromised. It must be kept in good repair, and this one had needed repairing properly for some time.

My boss' father measured the board and noted all the details he would need to order in a new board and shape it to suit the buggy.

I'm not sure of the exact quote, but let's say it was around $4,000. The owner of the buggy got quite a shock. He couldn't understand why the quote was so high.

My boss' father explained how this was the main structural member of the whole buggy and it must be kept in good order to ensure safety. He also insisted on replacing the whole thing, not just painting it.

The chap disappeared quite disgruntled and wishing he hadn't asked for the quote.

A few months later, they crossed paths again and the chap was pleased to announce that the whole job had been done for only $200.

My boss' father was shocked, saying, "I don't understand doing it for such a low amount."

The chap proudly announced, "I've been assured the job has been done properly and it even looks like a good job to me."

The shocked coach builder said, "Excuse me, but please allow me to test one corner of the new board; it won't take long."

"Go ahead if you don't believe me."

He happened to have in his pocket a small awl. This is like a small drill you can twist by hand to drill a small hole in timber.

He found a corner that was out of sight to people riding the buggy and drilled a small, discreet hole into the new barge board.

Remember, the barge board could be over fifty millimeters thick.

The awl was through the new structural member in about five millimeters.

The people who did the job had just placed a veneer of timber (a bit like plywood) over the board and glued it into place. They of course tidied up the edges so that the owner couldn't see the repair.

I don't remember now what took place after this but it would have been fairly terse and upsetting to say the least.

What is a 'wock'

It's a thing you throw at a wabbit when the wifle won't work.

White Fella Fishin'

I was invited to go fishing with my two mates so, being a typical white fella, I brought along some fishing line and bait. They didn't laugh at me 'cos they do this too; but they brought their special spears with them.

While standing or walking in the water's edge, they will wait until they see something and quite athletically launch a spear into the water.

Sometimes they wouldn't recover the spear but continue to watch what was going on under the waves.

But, if they threw the thing in, sometimes they would rush over and lift the spear up into air using the dry end – not the wet end. Whenever they did this there was always a small fish on the end of the spear wishing he was somewhere else.

> **White fella**
> an Aboriginal English word for non-Aboriginal person of European descent.

"Henry, how come you know when to run in and grab the spear and when not to?"

"You watch the end of the spear, silly. If it's wriggling, you've speared a fish. If it stands still, you've missed.

Speaking of spear fishing, I had to laugh one day when Henry said, "You know Doug, when we see you white fellas spear fishing like us, we have to laugh."

He was referring to the type of spearing where you walk into the water and look for a reasonable target. My mind tried to think of all the reasons this was wrong and landed on the idea that we were probably scaring away the fish.

"Why, Henry? What's so funny?"

"You white fellas don't wait until the sun is behind your head!"

Fair enough.
These guys know their stuff for sure.

Nan's Taxi Service

Somewhere the other side of Gunnedah, on that very rich soil country, lived a hard-working family. Mum and dad were the farmers, but children were expected to do their chores when they were older.

When you're feeding a family off what the land produces, you don't work nine to five. Dawn to dusk is more likely.

Farmer mums don't have a lot of time for running into town for shopping, picking up spare parts for the tractor or getting Gertrude to the dentist. Are you with me here?

This is where Nan comes in. Usually, mother's mother is called 'Nan' and grandad is called 'Pop'.

Over the years, Nan did most of the running in and out of town or to the co-op and back.

As you would expect, the three girls were teenagers now and they became aware of a problem with Nan's driving. You see, the maturing daughters became aware that Nan never had a driver's licence and were worried for Nan's sake.

"Nan, you must get your driving licence. If you have a prang, the car insurance will be no good. They'll refuse to help with the repairs."

The truth of what was being said was obvious to all, but Nan didn't fancy her chances at passing a driving test. Besides, she'd been driving all these years with no ... mishap, shall we call it.

The teenagers had their way, and eventually Nan agreed to, at least, go and apply for a learner's licence.

Next time she was in town, Nan reluctantly went to the police station (as we did in those days) and asked Sergeant Macgregor, "May I apply for a learner's driving licence please?"

The sergeant hesitated.

Mrs Smith looked at the floor expecting the officer's wrath.

"Mrs Smith, do you mean to tell me you don't have a driver's licence?"

A repentant "No," was her reply.

"You mean to say, you've been driving those lovely granddaughters of yours around all this time without a licence?"

"Yes, you see my daughter and hubby are flat out on the farm."

She was even more repentant now and fearing the worst.

He reached up to a pigeonhole and withdrew a document. Wordlessly, he filled it in and signed it.

Handing it to her he declared, "Mrs Smith, I've seen you driving around town for years now. Here's your full licence; there's no need for a driving test."

She nearly choked with joy and relief.

Can you imagine the family's joy as they passed Nan's driving licence around the table at teatime.

Try doing that in the 21st century.

Strange Things Happen in Aus. #2

A local farmer of many years would soldier on through wind, rain, snow and pain.

He must have been second or third generation and, by the time I met him, surely he would have had a few grandchildren.

Phillip was his eldest son and he did most of the farm work by now, but – by crikey – dad still did his fair share.

Try stopping these old guys from doing what they've done since their school days. I wish you luck.

But Dad had a problem, his old body was full of aches and pains (tell me about it!). To overcome his body's reluctance to working or moving he would take a painkiller powder with his breakfast every morning.

Over the recent years, he began to feel rather poorly and eventually heeded his wife's advice and went to see a doctor, endangering life and limb of everybody in the surgery. I jest of course but, believe you me, these old-timers were very, very reluctant to consult doctors.

It wasn't clear to the local doctor what might be causing the ailment, so he ordered some blood tests.

Dad went back to work – this time helping and supervising the shearing.

Several days later a phone call came. "Bring that old-timer in now, and I mean now!"

The farmer put in an appearance at the doctors.

"Sir, you have killed your liver!"

Apparently, this can happen if you continually take these old-fashion pain killer powders.

Dad sat down with a thud that tested the strength of the chair.

"Sir, don't worry. Your liver has sprouted again, you will live."

He did; but he had to learn to live without his favourite painkiller after breakfast each morning.

Strange Things Happen in Aus. #3

Following on from the dad who killed his liver, his son and daughter-in-law bought a farm out in the nearby bush.

Phillip had learned well from his dad and soon, he and Beverley had the farm up and running nicely.

By and by, they saw the need for a second income, and decided to create a fun park for the local and not-so-local families

As the years went by, their diligence saw the fun park increase and cages were built to house all sorts of native and unusual animals.

The reputation of the fun park grew and even schools would come for a day's outing. A food kiosk was opened and their popularity grew.

But, by now they needed more space for planned extensions. This concerned Beverley because there were old buildings and a lot of old (never to be used again) machinery in the way.

So the decision was made to auction off everything that would sell and clear the decks for improvements.

The day arrived, but the auctioneer was having trouble getting good prices. This didn't worry Bev as she was only interested in getting rid of the stuff.

"The next two items are tractors who had served previous generations well. What's my bid for these old, but well-known, make of tractors?"

It took his years of skill and practice to get to a lowly $250.

The successful bidders were probably as old as the tractors themselves. They fronted up to the desk where Bev was receiving payments and writing receipts.

They handed in the money, received their receipt and went to walk away.

"How do you plan to take the tractors away?" Asked Beverley anxious to help.

"Oh no, we don't want them. We just want to help you in your marvellous efforts out here."

It takes a lot to make a woman speechless, but she was, and it took her more than a few moments to recover.

We had a very unhappy Beverley on our hands – it was time for mice and men to retreat.

This did happen. True story.

Part 4: Morals

**When morals are in disbelief,
It's time to put things right.**

Beekeepers can be an Ingenious Lot

I'm guessing you haven't had much contact with Aussie beekeepers. Am I right or wrong here?

Well, they are an inventive lot. You have to be, it's a three-hour drive to a decent size town and even then they probably don't have what you need to fix a problem or invent a solution to a need. You get used to just using what you already have.

Okay, you can buy the bee boxes, but you want to make the floorboards and hive lid yourself. You've inherited dad's circular sawbench and the bunch of woodworking tools he had. They are now in your workshop that you built yourself and they are begging to be used.

There's no point just looking at them, they need to be used.

Okay, you mightn't be a genius, but you have inherited his skill and, what you lack in skill, you make up for in determination and mental aptitude.

We came to Australia in the late 1950's.

The local beekeepers were very welcoming and generous – nearly to a fault.

One of the locals also came from England in his early years.

He had a well-established bee business and a workshop that most of us would be jealous of.

But Peter was always looking to reduce costs and waste. This is a form of religion for beekeepers.

By the time we met him, he must have made hundreds of bee boxes, floors, lids and what have you.

One day, his inventive eye landed on the circular sawblade. Edge on, they can be two to three millimetres wide. Over the years, Peter grew tired of sweeping up and discarding so much saw dust after another session of making things using the sawbench.

So, Peter got to thinking – oh oh, that sounds dangerous, doesn't it?

IT WAS.

He unmounted the sawblade and got a local machine shop to cut the teeth off the half-metre diameter circular sawblade, leaving it perfectly round.

Next, putting his Aussie ingenuity to work, he drilled four equally distanced holes around the periphery of the blade. Each hole in a bit from the new edge of it.

"I'm not silly," said he.

This is the good bit.

In those days of limited refrigeration, bush people bought their fruit in cans.

Siezing an empty can, he cut the reinforced ends off and then cut the cylindrical remains lengthwise so he could flatten the thing.

Now with the can completely flat, he cut out four triangular pieces with a thirty-millimetre base.

Next, drilling a hole in the base of each one, he bolted the four teeth to the holes in the modified saw blade.

He admired his handiwork, realising each cut would now only be about 0.2 of a millimetre wide. Just think of the reduction in waste saw dust this was going to save!

During his first few trial runs, the thing worked beautifully. Okay, the depth of cut was now only about twenty millimetres but hey, the reduction in waste was fantastic.

Peter was yet to see the flaw in his plan.

It became obvious that the rotational speed of the modified blade was going to have to be increased. So, he had this ramped up a bit. A large bit actually.

Friends and fellow beekeepers were invited to see this beautiful invention.

He had a reasonable stack of timber – enough to make plenty of floorboards to impress his invited visitors.

They were impressed alright. When a small group had crowded around, Peter fired up his invention. They had never heard a saw spin that fast before. The howl of the sawblade made them all nervous and each took a step back.

They checked to make sure the shed side doors were open – you never know when the need to evacuate can happen.

This thing worked beautifully; they'd never seen a saw cut that narrow before.

Peter was in his element; the finished cut was better than anything seen before.

The demonstration continued and everybody was impressed. They found the howl of the modified sawblade disturbing, but the results spoke for themselves.

Around mid-afternoon, Peter's son Charles took over the demonstrations and was enjoying his newly found fondness of dad's invention. Not to mention his newly found popularity.

During one of his cuts, there was a sudden explosion from somewhere in the sawbench's internals. Charles staggered back in frightened alarm.

The piece of timber he was sawing was completely shattered.

Then at the same time, there was a second explosion in the shed roof.

The onlookers exited the shed in panicked disbelief.

Charles, on gathering his wits, quickly shut the power off. He stared at the shattered remains of his beautiful handiwork.

Then looking upwards, he couldn't believe the size of the hole in the roof. It was more than 0.2mm wide let me tell you.

Voices from somewhere else in the universe called out, "Are you trying to kill us, mate?"

Peter's get-rich-quick ambition instantly disappeared.

They went back to using the normal circular sawblades and a sawbench that didn't howl.

In case you're wondering what the explosions were, Peter had completely over-estimated (or didn't even consider) the tensile strength of his very thin fruit-can steel. He was lucky they lasted as long as they did.

When one of these teeth finally did break loose, it had enough force to partially demolish the sawbench internals *and* put a respectable hole in the roof iron. (I'm tempted to say Ozone layer)

Can you imagine the damage it would cause to human flesh!!!

The Town was in Chaos

The local radio station employed a morning show presenter that had an outrageous sense of humour coupled with a mischievous bent.

He was very popular, and had an audience far beyond the expected signal range of the station.

His humour was enjoyed by us all, however, some of his pranks were to be feared by local business houses.

His fondest wish would be to keep the Aussie spirit alive and well.

To say Terrance was popular with one and all (to the exclusion of his employers) was perfectly true.

There is one day of the year where it is quite acceptable to play a joke on folks or, as we used to say, to pull your leg.

I'm gonna let you figure out the date, as it has fallen into disuse today with everybody glued to their mobile phone or TV screen.

Oh for the old days ... dream on, Dougie.

Yes, that dreaded day of the year is here. Terrance had been planning this for weeks now. To say 'this was his crowning achievement' is probably untrue, but it might have been the trick he was most proud of.

To say the city fathers and local police were upset is a galactic understatement.

The 6am news, weather and sports report were concluded. Then instead of returning to the usual music and ads, we heard this URGENT MESSAGE from the town's police.

"A semi-trailer loaded with hundreds of sheep has overturned in the main street of town. Motorists are urged to avoid this area of town as the sheep are roaming around everywhere. Apparently, the driver has been extracted from the wreckage uninjured, but extreme caution is urged while the emergency crews are still there."

This message circulated around the farming communities with a speed that modern technology would be incapable of.

Remember the old reliable telephones? No upgrades, no credit limits, no recharging.

What happened??

Anyway, there are literally hundreds of farmers in this part of the state, each with at least one sheep dog.

Farmers can't just ignore news like this.

"Okay Ma, I'm heading in to help sort this out. It's time for us blokes to step up to the plate and serve our country!"

Looking back now and with many years of marriage experience, I reckon all the farming wives would have been a bit suspicious of this emergency – especially when you consider our women folk have a better idea of things than we do most of the time anyway.

Well, as you can imagine, hordes of over-anxious farmers, young and old, descended on the town.

Meanwhile, Terrance had been playing to air the sounds of hundreds of sheep with the accompanying farmyard sheep noises. Heaven alone knows where he got these from; he had prepared well – and it all sounded so authentic and convincing.

We now have a problem, and Terrance isn't feeling so happy with himself.

The problem: we have all these farmers and dogs searching for the wreckage and all the loose sheep they had been hearing about.

A lot of the local businesses were not impressed at the absence of patrons. What self-respecting women would dare come into town with all those farmers, sheep and sheepdogs roaming around freely.

And, after all, the radio did say to avoid the area!

I'm guessing by now the radio station's front door was securely locked.

Look here, a lot of what I have just portrayed is just from my imagination but this incident did happen. More than that I cannot say.

Let's just say the local constabulary were not impressed.

But it was the 1st April.

We Were All Young Once, Okay?

The Inverell PFA decided to hold a bonfire night at the Tingha member's place. Now I forget the occasion – maybe it was the Queen's Birthday weekend??

Resident farmer Graham had done well to have such a respectable pile of fallen timber ready for the occasion.

The fire was lit and we youngsters gathered around to enjoy the spectacle – not to mention the welcome heat on such a cold night.

There might have been three married couples, but most of us were still single – except for David and Chelsea who were engaged.

"Hope there's no-one home in those logs, Graham."

But there was, as we were about to find out!

The fire had a good hold by now and we were just thinking about going inside for drinks – the non-alcoholic kind, of course.

As is typical with mixed groups in those days and also with country custom – the boys were separate from the girls, and this was half the fun of what was about to unfold.

I'm guessing you don't know what a goanna is, am I right or wrong here? It is a usually long and fierce lizard with the sharpest nails (or claws if you will) and they know how to use them!

They can climb any tree at about half the speed of sound. Whatever you do, don't ever shake hands with one 'cos you'll lose a finger or some other part of your anatomy!!

This resident goanna got a touch upset when he felt his home get abnormally hot!

So here he comes. For starters, he wasn't happy when his log cabin was picked up and thrown onto a pile. Secondly, having his home transported away from his favourite stomping ground only added insult to injury.

Finally, this 'ere under-floor heating was just too much. "I'm gonna make my presence felt."

He came out of the fire, half-blinded by the smoke, and straight up the nearest tree. The tree happened to be David's right leg.

Permit me to introduce you to some reptilian thinking. When something wants to eat you, or just chase you for the fun of it, you head for the nearest tree and put those lovely sharp claws you have to good use.

Climb quickly, don't waste time looking behind you and don't stop until you reach a fork in the tree trunk.

Now you decide which fork will lead away from the danger and you go for it.

But wait ...

This stupid tree is upside down – I've reached the fork in the trunk but the other one leads down to the ground!!

"What the hell?" I'm gonna head upwards, but I'll have to try harder, there's some sort of restriction and it's trying to grab me.

Back to David who's in full panic mode. He has a goanna in panic mode in his crotch who is trying to get past his waist. This thing is using his claws like his life depended on it.

Meanwhile, David was grabbing the reptile and trying to stop his male furniture from being ripped to extinction.

In sheer desperation he yelled, "Chelsea!!"

Chelsea, still surrounded by her female friends responded, "Just what do you expect me to do darling?"

David, realising his future fatherhood was at stake here, retreated to the farmer's bathroom and dislodged the offending animal, then using towels, tried to mop up his bloodied legs and damaged male furniture.

Dad Wasn't So Sure

It was that time of year when farmers start sowing for the next crop.

But the weather was not being kind and the ground was too wet to sow a crop with the normal combine, so Dad had given up hope of sowing the twelve acres.

This was obviously a topic of discussion over the lunch table.

But, second youngest son was home on holidays from the Ag College in Armidale.

"Dad, we've been shown how to sow a crop on wet soil."

"Son, I've been farming since I was younger than you. Believe me, if you try to sow on wet ground: one, the tractor tyres chew the ground up making it even harder to sow, and two, the combine tynes continually block and you're forever stopping to clear them. No son, I very much doubt it and besides, sowing on ground that's as wet as the that block is, is a waste of time and effort."

"But dad, it can be done. Let me show you, please."

Dad got to thinking. Sometimes it's best to let youth have it's head.

"Okay, tomorrow we'll see this new method of yours."

Tomorrow arrived and we have a very light sprinkle of rain. Second son was not deterred, and the tractor and combine were hooked up. Dad was driving the small truck with the seed grain on it and we all make out to the twelve-acre paddock.

The combine hoppers are filled, the son fires up the old Massey Harris, and heads to the corner where we always start ploughing or sowing.

It's important here to know that the rate at which the seed is allowed to drop down into the ground would be something like three or four seeds for every ten centimetres or so, I don't really remember.

One row of tynes has a tube attached to it which ensures the seed gets into the ground behind the tyne and then the other tynes cover it up so it can germinate when it's ready to.

But, today we are using second son's new-found technology and the tynes are not in the ground, but are in the fully-retracted position. This means that the seed simply falls directly onto the ground and just lays there.

The son finishes the whole paddock and they both head home, he on the tractor-combine and Dad driving the seed truck.

Over the dinner table, Dad made a point of not mentioning the day's activities. At the same time, he imagined the birds would think all their Christmases had come at once. He also pictured all the flocks of birds coming to devour this bonanza of food.

It never happened.

Ag college holidays were over and son disappeared to continue his studies.

The rains eased and farming life continued.

A few weeks later, a light shower fell. Dad was starting to think he'd better go and re-sow the paddock normally. *I need that harvest.*

Dad opened the gate and walked into the paddock. "Good Grief."

He could not believe his eyes. There was a complete fence to fence carpet of newly germinated crop and he was standing in it.

"Holy smoke! That son of mine has taught me a lesson."

I never found out how well the crop did – because it was here that I left Dad's employ and started my apprenticeship.

But I'm betting quids; it was as good as any other crop sown in that twelve-acre paddock before.

Camping Mishaps

I was allowed to take our Boys' Brigade boys on a ten-day camp once a year.

This time I decided to spend the time at the Warrumbungles at Camp Wambelong. This is a wonderful place for young fellows from farming families or towns folk, to come and camp in tents and experience the lovely Aussie outdoors.

As usual, camp mum and dad were the bullockies of old who could produce beautiful food from camp ovens, billies, saucepans, and a simple BBQ plate.

Apart from hiking, outdoor games and just relaxing, I had brought along a heavy rope which I had strung between two trees.

I tied it on each side of the little creek, and then retied it about a metre and a half higher and brought it back and tied it off

---

Boys' Brigade
a Christian youth organisation
teaching boys about the
Christian faith, life and challenges

on the same tree I had started with. This allowed the boys to walk along the bottom rope and keep their balance by holding onto the top rope. Should you have an accident, plunge into the creek you went.

My boys' ages ranged between thirteen and sixteen. The rope bridge proved to be very popular with them and, after a few early mishaps, they acquired quite some skill at it.

However, boys being boys, they soon found walking the rope a bit boring, so they would have pillow fights out in the middle.

Two boys would arm themselves with a pillow or a bag filled with towels or shoes or whatever was around, and each starting from opposite ends, would meet in the middle and get stuck into it.

Okay, so we have some wet clothing and bodies, but these are easily dried in the normally warm weather of spring.

So far so good – but things were about to take a turn for the worse.

We're about halfway through the holidays when a bus load of boys from a Sydney Catholic High School parked in the camping area next to us. It didn't take them long to discover the rope bridge that I had built for us and could be heard using it with quite some excitement.

They thought it was there for all and sundry!

Our boys were disgusted because the newcomers were using it when they wanted a go. The other boys were slightly older and slightly bigger than ours.

Some of our boys would challenge a chap from the other camp – meet in the middle and have it out, shall we say! I could see trouble happening here, so I approached one the Catholic Priests and informed him that the rope bridge belonged to me. It's not there for public use. He was quite taken aback and became slightly agitated.

"Now you tell us!" said he.

Being of Irish Presbyterian descent myself I replied, "Look, I don't want another Catholic-Protestant war on my hands. We've only got a few days left before we head home to Inverell. Let's take this day about, shall we?"

He readily agreed and that was that. But I must say how proud I was when one of our younger boys won the slogging match and the bigger one fell into the drink. You see, our boys had several days to acquire some skill at this confrontation stuff, and it definitely showed. Our boys returned home, dry and warm, and boasted of their successes to their respective parents.

> ***The drink***
> the body of water, such as a dam;
> usually refers to someone or
> something being tossed into it.

Line Search Error

A young fellow was found missing from his bed one evening! (I'm not sure how one can be found missing, but that's what happened)

We figured out later that he followed mum's tail-lights down the driveway when she took the young fellow's friend home after visiting for the evening and then couldn't find his way back to the house – but anyway that realization came after the unsuccessful search.

We were called out at about 9.30 pm (we being a local search and rescue volunteer group) and arrived at the home to be part of a larger group to help find the youngster.

We were under police instruction and were told to search the machinery sheds and shearing sheds.

We searched under and over everything we could see in the dark. We searched the sheep yards, the gardens and more than I can remember – we took our job seriously.

Nothing found.

We heard a whistle and assumed we were wanted back at the homestead.

It was approaching midnight so the officer in charge gave his orders. "You rescue people, go home. Grab some rest and be back here before dawn."

We weren't happy, we were quite used to working into the night. We didn't let on but we retreated to our vehicles quite disgruntled. We went home, got some rest and returned to the homestead in very early dawn.

"You rescue people, I want youse to search the front paddocks between the house and the main road."

We did just that, we could see in the early light but still needed torches to see details. We lined up in the first paddock and Darren had the compass. He took a compass bearing and set off. So long as we kept level with Darren, and he kept the compass needle on the same reading, we could walk at a slow pace and be sure there was nothing on the ground that we overlooked.

About half a kilometre later, we reached the far end of the paddock. The end person stood still, and we pivoted around to line search on the way back.

The light was getting better but the compass was passed to me. I set the compass at 180 degrees to the first leg, and we set off.

All went well just as it had on the first leg. We reached the top of the paddock, the end person stood still, and we pivoted around and waited while Francis was handed the compass. He didn't need the compass as the light was getting better, so he took a fix on a distant object, and we set off.

We had only gone about fifty metres when he cried, "Stop, youse are all swinging round!"

"What's the matter, Francis? We didn't have this trouble on the first two legs. What's going on?"

We gathered around to enquire why we were having this spot of bother. The light was getting brighter by now so we asked, "What is the fix you took?"

"That rock over there."

We looked to see which rock he had chosen.

"Do you mean that brownish long one?"

"Francis, that's not a rock. That's a cow having breakfast!"

You see what was really happening was that Francis, on keeping his eye on the 'rock' in the distance, kept walking towards it – which is precisely what you're supposed to do – but it's not very often a rock will slowly move as it has it's breakfast.

This, according to Francis, caused the people on his right to gain ground compared to him and the people on his left to lose ground which made him think we were all swinging around.

We tried awful hard not to laugh.

Remember I said 'unsuccessful search'? Well, this is how things panned out.

It must have been late morning when two things happened at the same time.

The POLAIR helicopter was being prepared for a wider area search and, completely unknown to each other, the local chap who had turned up on horseback to assist in the search for the young fellow was about to set off and search the nearby TSR which came close to the house.

He mounted and navigated his way through the multiple yard gates that are always near the woolshed on a farm in the country.

Just then, the sound of a turbine engine wrecked the tranquillity of the environment – POLAIR was taking off.

Our friend on horseback had found his way onto the TSR and headed away from the house.

TSR
travelling stock reserves: Crown land set aside for grazing stock (horses and cows) while moving them around.

As I remember it, he was heading down the slope of the travelling stock route which can go for many kilometres in farm land.

It just so happened that the POLAIR chopper headed over the top of the TSR.

A little chap came out of the long grass (where he'd spent the night) and watched (as young folk do) the flight of the chopper.

This all happened in the path of the horse – also following the TSR.

The rider couldn't believe his eyes.

There, twenty or so metres in front of him, a little figure stared at the disappearing aircraft quite oblivious to the approaching horse. The rider got close, reached down and grabbed him by the shoulder saying, "We'll have you son!"

And that is how a lost little chap became a found little chap.

Our joy knew no bounds; not to mention how his mum and family felt.

Chopper
Helicopter

Signed, Fluor, Brown & Root

Our antihero of the moment (in the early seventies) was a janitor in a large workshop, situated at the bottom of a gorge where I worked.

The section of the long workshop where I worked as a motor mechanic/fitter was known as the 'light vehicle' workshop.

The trouble was, there was nowhere else for me to work on light vehicles and heavy haul truck engines and transmissions.

You can imagine the consternation of other heavy vehicle operators wanting to get past the back end of a haul truck as they needed to access a different section of the long workshop. And, especially at night, I often worked the graveyard shift.

There were three to five people working in this end of the workshop which included the light vehicle section, the machine shop, and the toilets. In the machine shop, Father ruled supreme. I never learnt his real name but no-one argued with Father.

It is in this world that our janitor of the moment worked. Chas was an awkward sod, could be very unpleasant to deal with, and very little in the world pleased him.

You could walk through from the light vehicle section into the machine shop and through into the toilets. This was Chas' domain.

But Chas was also the main 'go for' for the whole length of the building (which included the machine shop, light vehicle section, the spare parts store where Clive Brisset ruled with unquestioned authority, heavy vehicle shop, general purpose/air compressor section and finally the automotive electrical section).

From here you could walk through to the large blacksmith shop – and there lies another story for another day.

Anyway, back to Chas – ruler of the toilet domain and general 'go for'. Chas had a very sad character flaw, well known to his work colleagues (not friends) who worked with him for many years before coming to Copeton Dam. He had an unhealthy fondness of little children.

One day, a mum and dad with two very young children turned up at my section of the workshop. For what reason I don't know.

All work was abandoned in a frantic search for Chas before he saw the young family. He was found and assisted, shall we say, to a safe area away and out of sight to our unannounced visitors. This was achieved very quickly by the workers who knew him and would take no nonsense from him. He was kept in seclusion until the coast was clear.

Though Chas was a humourless chap, he did his job well and with a determination that brought great mirth to us all one day.

You see, our American bosses employed by Fluor, Brown and Root, worked on a monthly ordering system. Most parts, and all consumables, were ordered a month ahead. If stocks of an item got low or exhausted, it was a week or two or three before supplies would arrive and life came back to normal again.

It just so happened on this particular occasion that our American boss went to use the toilet for obvious reasons. On finishing his download, he reached for the roll of toilet paper to find there weren't enough sheets to do a satisfactory job.

He exited the cubicle in a foul mood and proceeded through the machine shop and through the light vehicle section on his way to his office above the spare parts store.

And who should cross his path, Chas. In threatening tones he said, "Chas, there's no toilet paper in the toilet."

"Yes boss, there's none in the store; we're out."

"I don't care what the excuse is! I want toilet paper in the toilet and NOW. Your job depends on it."

The boss climbed the stairs and disappeared into his office. Chas was still frozen to the spot. "Your job depends on it," the words echoed around in his head.

Said he to himself, "How the hell am I supposed to get fresh rolls of toilet paper? There's none here!! and it's not my b----y fault!"

Now, remember I was saying how Chas took his janitor job seriously and with admirable determination. It was probably the only admirable quality he had.

Chas disappeared off the planet for about an hour – not that anybody noticed; we just wanted to get on with our work.

Near the end of our shift, somebody came out of the toilet area laughing loudly.

"What's so funny?" he was asked.

"Have a look at our new type of toilet paper Chas has put in the racks behind each door."

Two blokes went in to investigate. They too came back with near eye-watering laughter.

In response to, "Your job depends on it," Chas had procured twenty or so sheets of emery paper, quartered them, and put the accurately squared smaller sheets in the racks.

Chas held onto his job but, no-one used the new style toilet paper.

Sometime during the night someone scrawled this notice on the end wall, in large letters for all to see:

> *New style toilet paper in use.*
> *Please use both sides.*
> *Signed*
>   *Fluor, Brown and Root.*

About the Author

Douglas Cornwall was born in Cornwall in south western Ireland and immigrated to Australia in 1958.

With experiences in youth ministry, and farming, the volunteer rescue association (VRA) and working as a mechanic, Douglas has lived and observed a *substantial* range of activities and adventures. But always through the eyes of the Irish larrikin – after all, that's who he is.

As honest and caring as a man in the ministry, yet as ocker as an Aussie on the land, Doug shares stories (most of them true) about some of the people who made him laugh and cry.

Strong and playful wearing overalls or greens, his yarns will have you seeing life and all it offers through a fresh lens. Just like seeing a crow flying backwards.

Doug now enjoys time helping his community as a volunteer driver and telling jokes to his grandchildren. And recording his many observations and yarns.

About the Artist

Leanne Deering is an illustrator and graphic designer known for creating book cover artwork and multimedia designs. Through a blend of digital and traditional methods she brings her hand-drawn illustrations to life.

Leanne lives in Newcastle Australia and has a double degree in the creative industries. When not immersed in her design projects she enjoys drawing, photography, reading books and watching movies.

Acknowledgments

Several of the stories and jokes in this anthology can be attributed to other story tellers. Sadly, some have unknown origins but would like to thank the originators for bringing joy into the world through their words.

I wish to thank acquaintances and family for the following contributions and inspirations:
* Camping Mishaps
* Can We Help You?
* Christmas as Cheesering
* Dad's New Tractor
* Dad's Old Station Wagon
* The Bendemeer Bridge
* The Butcher's Name was Mr Push
* We Were All Young Once

The following are from unknown sources:
* I'm Nott the Rabbit Inspector
* Nan's Taxi Service

And the rest? Maybe, just maybe, they're true stories, written by Douglas Cornwall.

Aussie lingo for the newcomer

Not all of these are still commonly used today, but were relevant at the time of the story

Argy bargy – argumentative talk, wrangling

Artesian - underground water supply

Aussie – a person from Australia (not often used to describe the country of Australia).

Belly up – to go broke

Boys' Brigade – a Christian youth organisation teaching boys about the Christian faith, life and challenges.

Bullockies - those who drive the bullocks (a castrated male of a bovine animal)

By jingoes – an exclamation expressing agreement.

Chopper - helicopter

Coo'ee – a signal used in the bush to attract attention. The first syllable is elongated and the second is of a higher pitch.

Crikey – an expression of surprise

Cuppa – a cup of tea or coffee

Downunder – relating to Australia or New Zealand

Dummy spit – a childish display of exasperation or bad temper.

Dunno – a lazy way of saying "don't know"

Dunny – an outside toilet, usually away from the house

Farmhand – a person hired to work on a farm

Freight on – the cost of shipping / postage / transportation has been added

Gonna or gunna – A lazy way to say "going to". A person who intends to do something but never gets around to it.

Gotta – A lazy way of saying "got to" as in "I've gotta go now".

Jigger – someone who jigs (a type of dance). It can also refer to just about any mechanical equipment that jolts.

K's or k's – kilometres, the Australian version of measuring distance.

Long-drop – a pit toilet

Lube bay – the area in a mechanical workshop used for lubricating vehicles.

Menfolk – the male members of a community or family.

Mozzies – mosquitoes, a much-hated insect.

Nuffin' – A lazy way of saying "nothing".

Outa – A lazy way of saying "Out of", such as "I'm outa here"; not to be confused with outer which means the outside of something.

Pick off – select one by one.

PMG - The Postmaster-General's Department provided the postal and telegraphic services throughout Australia.

Pothole – a hole in a road

Probs – A lazy way of saying "probably" or "problem"

Struth – an exclamation expressing surprise or verification.

Techie – A lazy way of saying "Technician" which is someone with expert knowledge on the technical elements of a subject.

Telecom – Australia's earlier telecommunications systems.

The drink – the body of water, such as a dam; usually refers to someone or something being tossed into it.

Thingy – an object or event which the speaker can not (or does not wish to) identify.

TSR – travelling stock reserves: Crown land set aside for grazing stock (horses and cows) around.

Wet season – the period of heavy rainfall, usually October to April in the northern parts of Australia.

Whatnot – an insignificant item

White fella – an Aboriginal English word for non-Aboriginal person of European descent.

Whompa - a dense sound made by machinery such as helicopter blades

Yeah – A lazy way of saying "yes".

Youse, yous or y'all – single or plural word for referring to another person or persons. "Can yous all help me with this?"

And ...

Can birds really fly backwards?

Ask Doug, he'll gladly

tell you the story.

www.ingramcontent.com/pod-product-compliance
Lightning Source LLC
Chambersburg PA
CBHW022013290426
44109CB00015B/1162